Wild-Caught Discus

by
Bernd Degen

Translated by:
U. Erich Friese
General Curator
Sydney Aquarium
Sydney, Australia

Dr. Herbert R. Axelrod (left), the discoverer of most of the color varieties, species and sub species of discus, and the author in 1990.

Bernd Degen

Wild-Caught Discus

I would like to take this opportunity to thank all the discus fans who have helped to produce this book.

My particular thanks go out to the following individuals, who have contributed photographs:

Front cover: W. Junge
Dr. Herbert R. Axelrod
Teoh Mee Ming
Werner Colle
F. Bodenmüller
Dr. Schmidt-Focke
Helge Musstopf
T.F.H. Publications
Hub. Kleykers
Dr. Wolfgang Staeck
Y. Nakamura
Gerd Thierbach
Werner Dallwitz
Martin Trostel
O. Richter
Fumitoshi Mori
Aqua Magazine Japan
Minuro Matsuzaka
Gan, Singapore
Transfish, Munich
Mr. Werner
Mr. Mahr
Aquarium Glaser, Weiskirchen

© 1993 bede Verlag GmbH for original German text.
© 1995 by T.F.H. Publications, Inc. for English edition.

All rights reserved. Not liable for any damages arising from implementing any text, data or other relevant detail presented in this book.

Published by T.F.H. Publications Inc., One T.F.H. Plaza, Neptune City, NJ 07753

Table of Contents

Foreword 7
Discus Classification 8
   Symphysodon discus Heckel, 1840 8
   Symphysodon discus discus 8
   Symphysodon discus willischwartzi 11
   Symphysodon aequifasciatus Pellegrin, 1903 12
   Symphysodon aequifasciatus axelrodi 16
   Symphysodon aequifasciatus haraldi 27
   Symphysodon aequifasciatus aequifasciatus 32
Importing Discus 38
Selecting Wild-caught Discus 53
Quarantine Treatment 59
Setting up a Discus Tank 63
   Filtration 63
   The Correct Water 66
   Total Desalination 69
   Reverse Osmosis 69
Care and Breeding of Wild-caught Discus 72
   Correct Rearing Methods 80
From Egg to Juvenile 85
Genetics 86
   Mutations 89
   Hybridization 90
Cross-bred Varieties 94

Foreword

Wild-caught discus are the real kings of Amazonia for us aquarists. After their discovery and eventual importation, keeping them in aquaria was initially reserved for a few specialists. Years later these people succeeded in producing magnificent captive-bred discus from the initial wild-caught stock. Eventually, wild-caught fish were replaced in the tanks of discus fans by their turquoise colored, captive-bred progeny. It took more than 20 years until wild-caught discus grew in popularity again. Now a high point has been reached, which is supported by many publications. Now one finally seems to again remember and appreciate the incomparable beauty of these majestic fishes. There has been a worldwide increase in demand for these wild-caught fish. Let us hope that this will not lead to increased habitat destruction so that these fish will continue to thrive in their native waters. This is an appeal to responsible breeders, to breed wild-caught discus in their natural forms and to keep the species pure.

Crossing wild-caught stock with all sorts of captive-bred progeny is already going on worldwide, so that new color varieties are constantly being developed, which are quickly picked up by aquarists who are purely interested in something "new." But in spite of that the preservation of purebred, wild-caught discus in our aquaria must not be neglected.

Bernd Degen

CLASSIFICATION OF DISCUS

Discus are members of the family Cichlidae. They are native to the Amazon River system, and its tributaries, in Brazil, South America. Discus are only found in that part of the world. They have gotten their name from their typical round (disc-like) body shape. There is some confusion within the tropical fish trade in regard to the scientific classification of wild-caught discus. Although only five wild-caught species and subspecies are involved—a relatively small number by any standard—it is not easy to point out the specific differences between them. In the final analysis, page-long treatises about individual species and subspecies with counts of scales and rows of scales has led to confusion among aquarists. The traditional wild-caught discus varieties are placed into the following five categories:

1. *Symphysodon discus discus* Heckel, 1840 – Heckel's Discus
2. *S. d. willischwartzi* Burgess, 1981 – Willi's Discus
3. *S. aequifasciatus axelrodi* Schultz, – 1960 Brown Discus
4. *S. a. aequifasciatus* Pellegrin, 1904 – Green Discus
5. *S. a. haraldi* Schultz, 1960 – Blue Discus

Latest findings seem to suggest that there are actually only two wild-caught species, *Symphysodon discus* Heckel and *S. aequifasciatus* Pellegrin, the latter occurring in many color varieties. These discus tend to hybridize among one another, and consequently differently colored discus are caught in various river systems that all belong to *Symphysodon aequifaciatus*. Different shades of color simply do not make new species. In order to clarify the traditional color designations, they are being maintained in this book. Importers and dealers still adhere to this system.

SYMPHYSODON DISCUS HECKEL, 1840

Symphysodon discus discus Heckel, 1840

In 1840, Johann Jacob Heckel, an ichthyologist and systematist at the Natural History Museum of Vienna, described a discus fish for the first time. The fish came from the Rio Negro region and was described as *"Symphysodon discus."*

The first live, wild-caught Heckel Discus were imported to the United States around 1930. American aquarists called them "pompadour discus," and they were sold at enormously high prices. In Europe the first Heckel Discus arrived about 1958, and in the 1960's the first Heckel Discus started to appear in German aquaria. Captive breeding of this species has been—and still is—very difficult and seems to succeed only occasionally.

Symphysodon d. discus Heckel is usually caught in the Rio Negro region and at the mouth of the Rio Branco. The exporters of this species are located in Manaus, the center of Amazonia.

The Heckel Discus can easily be recognized by its prominent cross bands. The fifth (middle) band is particularly well-defined and clearly visible. This is not the case in the other species. Similarly, the first band crossing the eye and the ninth band through the caudal peduncle are more strongly indicated than in the other discus species. Depending upon the mood of the individual specimen, the Heckel Discus displays these bands either strongly or weakly.

Symphysodon discus—The species discovered in 1840 by Heckel. The Heckel Discus can be clearly recognized by its strongly developed central band. Other distinct bands run through the eye and caudal peduncle. These characteristic band markings make the Heckel Discus distinctive. For that reason S. discus must be considered a valid species. Depending on the collecting location, Heckel Discus display some variation in their basic coloration.

→ *Rarely do we see such an intensely blue base coloration in Heckel Discus. If only the head had bright blue colors, it would be referred to as "blue-headed Heckel." Unfortunately, such magnificent specimens are rarely ever available. The discus shown here appears to be well settled in, because its black stripes are only barely visible.*

↑ *Heckel Discus with brown base coloration and delicate bluish stripes are common. In this specimen from The Rio Negro the deep red eye is rather conspicuous.*

→ *Heckel Discus occur in the The Rio Negro, which is a typical blackwater river. Its water is distinctly brownish. The pH value is usually below 4.0.*

Since Heckel Discus occur in the black waters of the Rio Negro, this fact must be taken into account when keeping them in captivity. The water chemistry values of the Rio Negro and Rio Branco must be considered as being extreme. The pH usually varies between 3.2 and 4.5, and will rarely rise above 5. The conductivity of this water is also extremely low. Values of around 10 µS are commonly observed. The water has a brownish color, which is indicative of a high content of decomposing plant material. Because of the decaying plant matter, the water is strongly acidic, which explains the low pH value.

Keeping Heckel Discus in captivity was initially rather difficult, since aquarists knew little about the water chemistry in their natural habitat. This was the reason why Heckel Discus have always been described as being difficult to keep. Since the water chemistry required by this fish has now been determined fairly accurately, it is, of course, much easier to keep. Because of modern technology and water conditioning methods it is very easy to provide the correct water conditions for this species. Since the pH is very low in the Rio Negro, the availability of food in these rivers is limited. Because of the strong acidity of these blackwater rivers, it is difficult for mosquito larvae to survive there. This is also the reason why blackwater rivers are pleasant travel destinations for tourists since there are hardly any mosquitos.

Breeding the Heckel Discus is, to this day, still rather difficult. In fact, it has been categorized as the discus most difficult to breed. So far only a few serious discus specialists have been able to breed Heckel Discus successfully. All other wild-caught discus are being bred in large numbers.

Symphysodon discus willischwartzi, Burgess, 1981

The American ichthyologist and tropical fish specialist, Dr. Warren E. Burgess, described this Heckel Discus subspecies in 1981. It was discovered by Dr. Herbert R. Axelrod in the Rio Abacaxis. It derived its subspecies name from Willi

Schwartz, the well-known discus pioneer and exporter from Manaus. Schwartz was one of the most significant discus exporters in Manaus. This discus variant could possibly be a natural cross between *Symphysodon discus* and *S. aequifasciatus axelrodi*. The description of this new subspecies of *S. discus* is based on a larger number of rows of scales between the operculum and the caudal base. But in appearance *S. d. willischwartzi* is more reminiscent of a typical Heckel Discus with its broad central band.

Another Heckel Discus variant was discovered and described by the American Dr. Herbert R. Axelrod; this is the so-called "Cabeza Azul Discus," which was caught in the Rio Jau. It is characterized by an intensively blue head and a high component of red over the body. Unfortunately, importing this rare discus is only infrequently successful, because the Rio Jau is a preserve.

SYMPHYSODON AEQUIFASCIATUS PELLEGRIN, 1903

The French ichthyologist Jacques Pellegrin described this fish as a new variety of the species originally described by Heckel. In his description, Pellegrin pointed out that those discus specimens originating from around Tefe and Santarem displayed significant differences from the typical Rio Negro discus. *S. aequifasciatus* was eventually established as a distinct species.

In 1960 the American ichthyologist Dr. Leonard P. Schultz published a treatise on the genus *Symphysodon*, in which he provided a valid listing

← *A recently collected blue-headed Heckel (in net) displays particularly conspicuous blue colors in the head region. The nicely rounded shape and the deep red eye make this wild-caught discus a prize specimen. Please also note the conspicuous straight line pattern along the sides of the body, which is typical for all wild-caught Heckel Discus.*

↑ *Heckel Discus are often imported in medium and small sizes by wholesalers. This gives an aquarist the opportunity to raise his own fish. But with that comes the risk that the fish may not grow up normally, and the body develops a marginally elongated and pointed appearance. Rearing small, wild-caught specimens is not easy.*

←*The discus importer maintains his fish in large tanks, where a number of specimens can be kept together. At that point it is already important that the somewhat more delicate Heckel Discus be given optimum water quality conditions.*

Not all Heckel Discus have an optimal pattern of longitudinal stripes. In this specimen the posterior part of the body no longer carries them.

This specimen has a more brownish (almost golden-colored; typical!) base coloration. The fifth band over the body is only weakly developed. It was collected in the Rio Curuim.

Heckel Discus are water and food specialists. It can take months until fully grown specimens will properly settle in. This may require considerable patience on the part of the aquarist!

Typical Heckel Discus caught at the Rio Negro, with well–developed blue colors and a reddish-brown base coloration. On this specimen, too, the deep red eye is conspicuous, which in Heckel Discus is not always the case. A majority of Heckel Discus have an amber-colored or yellow eye. In this magnificent specimen the upper edge of the dorsal fin has also attractive red coloration. Generally, wild-caught discus do not get as large as their captive-bred progeny (reared under optimum conditions!). The normal size range for adult wild–caught discus is 14 to 16 cm, while their captive-bred progeny can reach adult sizes of 18 to 20 cm.

of the species. According to Dr. Schultz three wild subspecies of *Symphysodon aequifasciatus* can be distinguished:

1. *Symphysodon aequifasciatus aequifasciatus* Pellegrin, 1903

2. *Symphysodon aequifasciatus axelrodi* Schultz, 1960

3. *Symphysodon aequifasciatus haraldi* Schultz, 1960

Symphysodon aequifasciatus axelrodi Schultz, 1960

Dr. Herbert R. Axelrod, a well-known discus specialist, has collected many Brown Discus specimens in the proximity of Belem (Brazil). And many of these fish are collected these days in waters around Santarem and Alenquer, as well as in the Rio Tocantins, Rio Tapajos and Rio Xingu.

Especially during the last few years, the Brown Discus have become very popular again, because certain color varieties were discovered which displayed an intensive red coloration. For instance, Dr. Schmidt-Focke received a shipment of wild-caught discus which had been collected around the town of Alenquer. The females of this variant displayed an intense red color and maintained this coloration while being kept in aquaria. The males, however, had narrow green bands, which in some specimens were distributed over the entire body. One pair of these specimens bred successfully in captivity.

The basic coloration of *Symphysodon aequifasciatus axelrodi* is light yellow to dark brown. The body is covered with nine vertical bands. The first one (just as in all other discus species) runs through the eye. The first and last band are somewhat better defined. This wild-caught discus became established in the tanks of many aquarists as the Brown Discus, so that during the 1960's and 70's it became the most commonly kept discus among aquarists. It could be said that with it, the discus fever broke out. These fish have

For years Brown Discus were absent from the tanks of hobbyists. Now they are very popular again.

A pair of Brown Discus can quickly become the esthetic centerpiece of a planted aquarium. It is the color contrast that provides the fascination.

Green Discus from the Rio Tefe.

Green Discus from the Rio Tefe.

← Brown Discus (of unknown origin) from the tanks of Dr. Schmidt-Focke. The well-delineated center band, reminiscent of Heckel Discus, is conspicuous. These fish have been imported as individual specimens only.

some bluish, contrasting stripes in the head and abdominal region, which give it a very interesting appearance. With a proper diet and optimum water conditions, the Brown Discus displays an intense rust-brown coloration, which makes this fish even more interesting.

The Brown Discus was also the first wild-caught type of discus which was easy to breed. Later on Brown Discus were displaced by green and turquoise-colored discus.

← Brown Discus, presumedly from around the Alenquer region. Conspicuous here is the fact that the female tends to show a stronger reddish coloration. The male's horizontal stripes are bluer, especially in the abdominal and head regions.

Brown Discus

Brown Discus

Plainly colored wild-caught Brown Discus, nearly evenly colored and without blue body stripes. ↓

↑ *Probably a captive-bred cross between a Brown Discus and a bluish-green discus. The indistinct, somewhat washed-out base coloration of the body makes it impossible to categorize this specimen. But the body shape, finnage and red eye are very good.*

→ *Wild-caught Brown Discus with an even base coloration, presumedly from the area around Santarem. Brown Discus with a lighter base coloration are often found in that area.*

Young, wild-caught Brown Discus with light golden–brownish base coloration, as is typical for fish from the area around Santarem. Characteristic again is the absence of light blue stripes along the body. The red eye is dominant. Brown Discus will breed more readily than the other varieties. Nobody knows why this is so!

22

← *An increasing number of discus have been imported from the area around Alenquer in recent years. Conspicuous in most of these fish is the deep brown-red base coloration. Females usually do not show any significant pattern of horizontal stripes. Males have more well-defined markings. Dr. Schmidt-Focke was the first to breed a pair of these fish.*

↓ *Brown Discus from the area around Santarem with light brown base coloration and a delicate pattern of blue strips in the head region. Also conspicuous is the deep red in the fins.*

← *All of these wild–caught discus were imported simultaneously from the area around Alenquer. The specimen in the foreground is conspicuous due to its high component of turquoise coloration.*

→ *In captive-bred Alenquer discus many of the progeny in the F_1 generation already show a deeper red coloring in the outer body areas than their parents.*

↑ *A pair of Alenquer Discus from the tanks of Dr. Schmidt-Focke. The female on the left has an intense brown-red base coloration. The male in the back is colored completely differently and could be considered – from a more classic point of view – as a "Blue Discus."*

→ *Progeny of Alenquer Discus. These specimens also exhibit an attractive intensive brown-red coloration. Moreover, the deep red eye and the perfect body shape make this a magnificent fish. The Alenquer Discus cannot, of course, be a new subspecies of discus.*

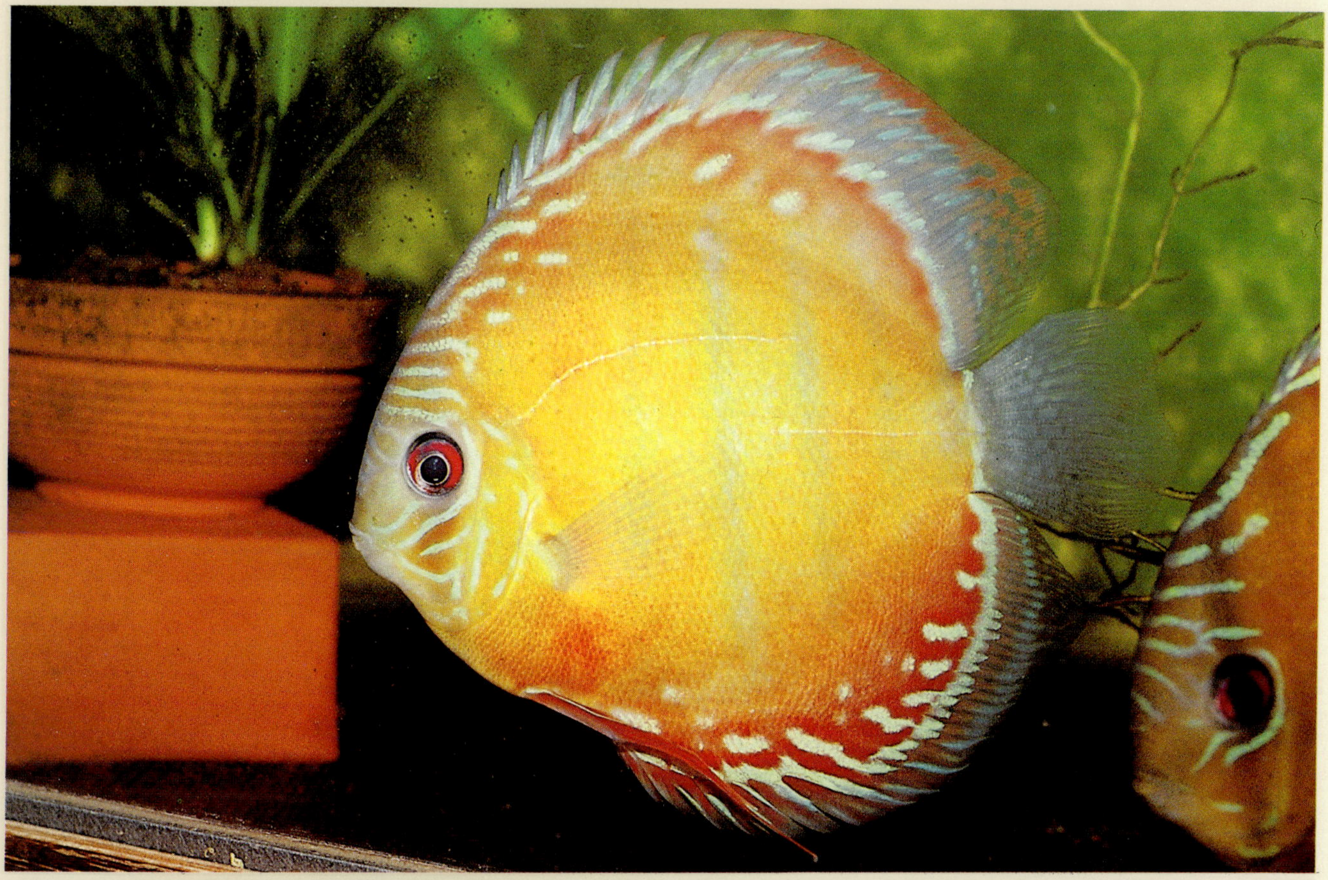

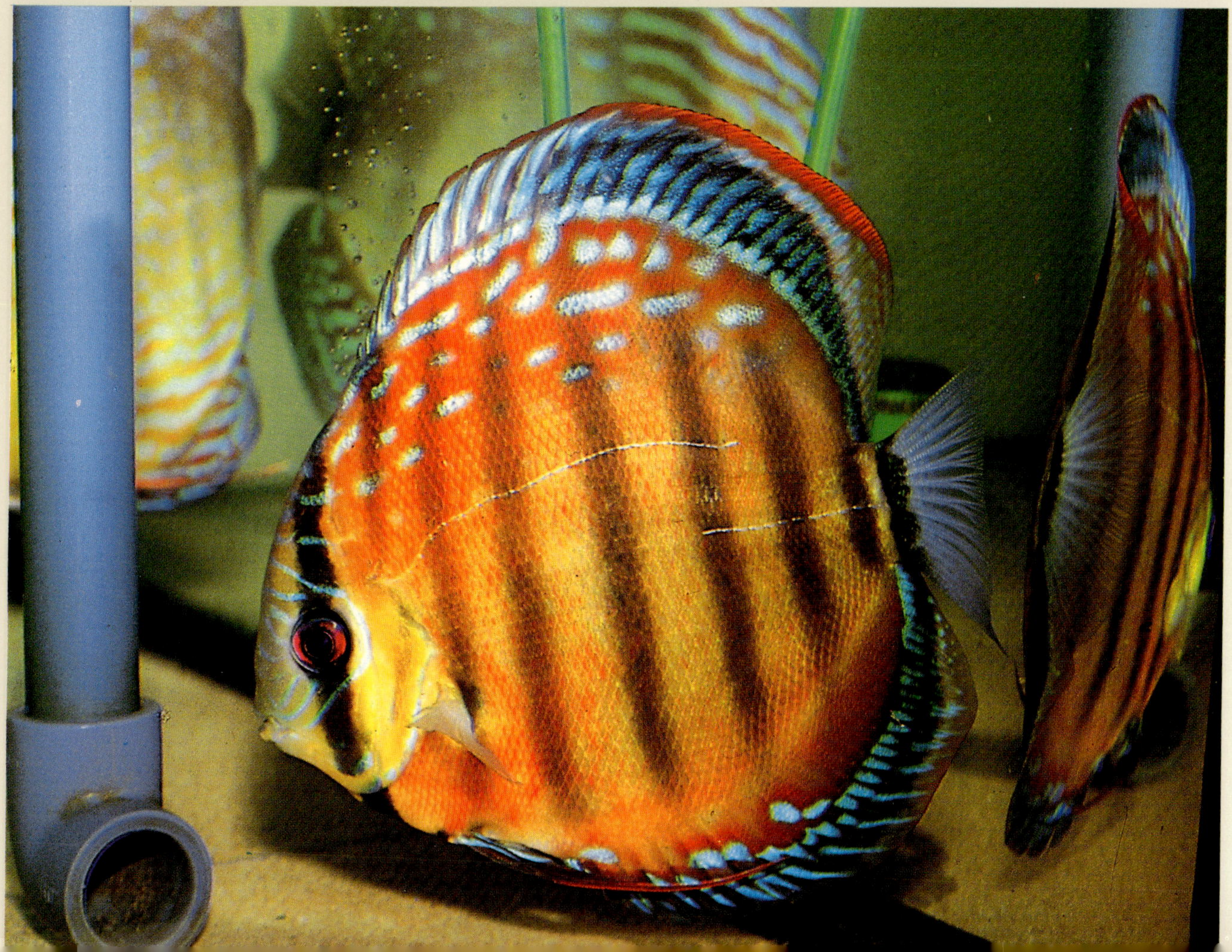

Symphysodon aequifasciatus haraldi is the classical designation for the Blue Discus. According to the latest systematic research, S. aequifasciatus is the name which embraces all color varieties. Coloration and markings among discus do not justify establishing new subspecies. In my opinion the designations Blue Discus, Brown Discus and Green Discus appear to be scientifically outdated. Nevertheless, it may take years until this is accepted by aquarists. This appears virtually impossible in the Asiatic region, where there are dozens of trade names for wild-caught fish as well as for captive-bred fish.

Symphysodon aequifasciatus haraldi Schultz, 1960

This discus is referred to as the Blue Discus. Harald Schultz caught the first specimens of this subspecies in the three-country triangle of Peru, Columbia and Brazil. The basic coloration of the anterior half of the body is brownish and becomes darker toward the tail. The head is purple and there are nine dark brown vertical bands on the body. Again, the first one and last one are more conspicuously dark. There are also intense blue stripes along the body, especially on the head and in the pelvic fin region. They are spaced in an uneven pattern over the body. The overall appearance of this discus is more colorful, with the actual pattern being better defined.

Those specimens covered all over with blue stripes are referred to as Royal Blue Discus.

The first Blue Discus were imported into Germany toward the end of the 1950's. In 1961 Dr. Schmidt-Focke succeeded for the first time in breeding these magnificent fish. He had received his first specimens from Harald Schultz. Distinguishing between Blue and Green Discus is sometimes not easy, because distinct color markings can only be defined with difficulty. Blue Discus are exported throughout the world from the Rio Purus and Lake Manacapuru region. These magnificently colored discus have given rise to superbly colored varieties by crossing them with green and turquoise-colored discus. Many discus varieties have arisen from Blue Discus.

Common Blue Discus are not fully marked. The specimen depicted above is a normal Blue Discus. The differences from the Royal Blue Discus on the opposite page (page 26) are obvious.

Royal Blue Discus

↑ *Royal Blue Discus from the Rio Manacapuru. Only a few specimens with a complete striped pattern ever reach the trade. They are always designated as Royal Blue. Wild-caught specimens from the Rio Manacapuru display a distinct brownish-red base coloration covered by straight turquoise–blue longitudinal stripes. The bright red eye is also conspicuous. There is nearly always bright red in the fin margins. Captive-bred progeny tend to develop full coloration early, and after nine months the foundation of an intense color development becomes obvious.*

← *Discus from Rio Manacapuru also often display a very intense brown-red coloration in the head region. Especially the combination of a red-brown body color and the bright turquoise-blue stripes make these fish absolute favorites among aquarists. The brightly colored specimens appear to be lead animals, which – logically – occur only in small numbers in the wild.*

Three photographs of discus from the Rio Purus. In these Blue Discus the black band, which extends from the base of the dorsal fin through the caudal base and on to the origin of the anal fin, is well developed. However, other discus also display this characteristic band, especially wild-caught Green Discus. This is more proof that all discus will breed with each other in the wild, and that a definitive categorization is actually not possible. The base coloration of these Blue Discus is always an intensive red-brown, whereby the brown color tones tend to dominate. The turquoise-blue striped pattern is variably strongly developed.

The two photos above depict the same specimen. The differences in coloration are caused by different lighting methods. The specimen shown is a relatively young, well-developed wild-caught discus from the Rio Purus. The relationship between eye size and body size is optimal.

This magnificent wild-caught Royal Blue Discus was collected in a tributary of the Rio Purus. The striped pattern is irregular and frequently interrupted. For wild-caught Royal Blues there is an emphasis on getting fish with an evenly spaced pattern of blue stripes. In captive-bred progeny this characteristic can then be selectively bred for. Nevertheless, this discus should be categorized as a high-quality wild-caught specimen. Sex determination among these fish is not possible. One can only make assumptions, which can turn out to be totally wrong. Wild-caught discus require a long time to become acclimated. Up to two years may go by before some specimens will mate and breed.

Green Discus from the Tefe region often display a weak red dotted pattern along the lower half of the body. The number of dots involved is, however, subject to considerable variation.

Symphysodon aequifasciatus aequifasciatus Pellegrin, 1903

Pellegrin examined the specimens caught at Lago Tefe. These had a brownish green base coloration and nine vertical bands of dark brownish appearance. All of the bands were of identical intensity. Since the dark vertical bands are in the same position as in the other discus species there were no obvious differences. The dorsal and anal fins are dark with a black margin. The dark, almost black coloration of the fins' edge is typical for this discus variety. There are also dark brown and iridescent greenish alternating horizontal stripes on the head and dorsal regions as well as in the abdominal region. The greenish-turquoise colored striped pattern is especially characteristic in the head region. Overall, these fishes appear to be somewhat "greener," which may have led to the name Green Discus.

The Green Discus is the classic third subspecies of Symphysodon aequifasciatus. *It is optically distinctive from its blue and brown relatives. Most of these Green Discus originate from the Rio Tefe or Lago Tefe. In general, all Green Discus have a very dark, black fin margin.*

Green Discus with a red dotted pattern that extends over large parts of their bodies, are allegedly caught in Lago Coari. Whether this is indeed a variety is doubtful. Presumedly the more or less abrupt break in brown-red body coloration has occurred in crosses between color varieties.

In this Green Discus the strongly developed brownish base coloration is very obvious. Of interest are also the bright red dots, which are very well developed. The turquoise coloration in the lower fin margin is strongly developed and is interrupted only by a few red dots and stripes.

A young Green Discus, about 8 months old, with colors that are not yet fully developed. Full coloration can not be expected until the age of 12 to 14 months.

Young Green Discus from Lago Tefe without any signs of red dots. The turquoise-colored stripes in the head region are only ill-defined. With increasing age these stripes become more distinct. Overall, it can be stated that plain-colored young wild-caught Green Discus will become more colorful as they get older. ↓

↑ This young discus also is not fully colored yet. It will certainly surprise its owner later on with excellent coloration. In the middle of the body there are already delicate lines of red bands visible. Based on my own experience I can strongly advise you to rear young, wild-caught discus, because they will pay you back later with magnificent coloration.

← In spite of its juvenile age this young Tefe Discus already shows the onset of excellent coloration. In time the red coloration will become more intense yet. This fish has an immaculate body shape and promises to turn into a superb wild discus. Waiting will definitely pay off with this specimen.

Green Discus – this color variety used to be referred to as the Pellegrin Discus.

↑ *Top: Pair of wild-caught Green Discus from the Rio Tefe. The male on the left has a more intense basic green coloration. On the other hand, the female on the right displays a distinctly brown base coloration. Once again, both fish show the deep black fin margin. Also conspicuous is the bright red eye. Both specimens are showing "fright coloration" with their nine, well-defined vertical bands. Once the fish have settled down again the black bands will become less intense.*

← *Newly imported, wild-caught Green Discus. Specimens with such a pattern of continuous green stripes are often traded as "Royal Green." The body shape can be considered as optimal, because this specimen really has a circular body shape. Minor skin damage heals promptly under proper care and should not be a reason to reject specimens like this.*

Importing Discus

Discus are native to Brazil. The vast Amazon river system is their home. Although the natural distribution of these fishes includes small parts of Peru, Columbia and Venezuela, Brazil is the principal exporting country of wild–caught discus. The gigantic rainforest region of the Amazon basin with its many large rivers is beyond our imagination. Anyone who has ever had the opportunity to look for discus on the Rio Negro will retain these impressions for the rest of his or her life. Most tributaries of the Amazon originate in the Andes Mountains. Depending upon the type of water carried, one distinguishes three types of river: white water, black water and clear water rivers. The best-known and largest are the white water rivers of Amazonia, with the Solimoes, Rio Branco and Rio Madeira. The more significant clear water rivers are the Rio Tapajos and Rio Xingu. And indeed, without doubt, the best–known black water river generally is the Rio Negro.

The white water rivers of the Amazon region carry a cloudy, clay-colored water permitting little visibility below the surface. From an optical point of view white water rivers are essentially opaque. This type of water carries a large load of sediments, which are being transported to Amazonia where they are deposited. This floating, alluvial material is transported from the Andes Mountains to the Atlantic Ocean. The pH value of white water is almost neutral, in fact barely under the neutral mark of 7. The electrical conductivity of this water is about 30 to 60µS, which is still relatively high for Amazonian conditions. Because of the vast amounts of sedimentary matter that is carried along, the river banks are constantly undergoing changes. These changes are most clearly visible during the rainy season.

Top: Newly caught discus are kept in nets which are simply suspended in the river. There the fish are held by the native fishermen until a buyer comes and takes them. This keeps the fish in good shape.

Bottom (right): Boats like this with their spartan fittings often accommodate the entire family, which goes along on collecting trips. In fact, most of the daily life of the family takes place on board such vessels. They are living room, bedroom and workplace all at the same time. Because these are shallow-draft vessels, they can easily navigate the river systems and catch fishes, even at low water.

During the dry season the water level drops several meters in the river.

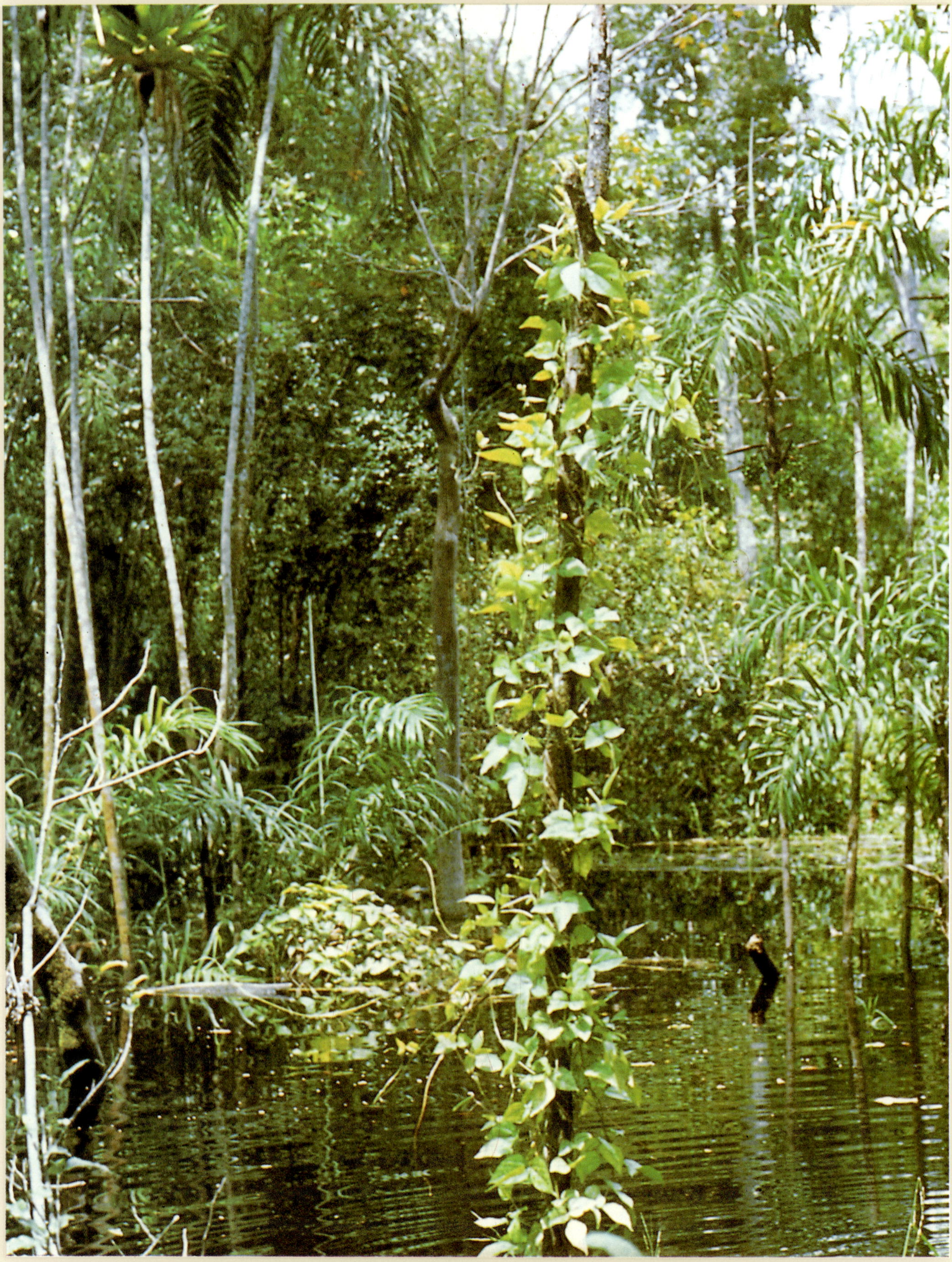

← *Caption for photo on page 40: The flood regions of the Amazon River system accommodate enormous masses of fishes. It is, however, difficult to navigate on these waters with boats because of numerous submerged branches, tree roots and other obstacles. The propellers of the boats keep getting caught in submerged thickets. It is easy to collect fishes there at night. By shining a spotlight on them they remain motionless and can then be simply scooped-up with a hand net.*

↓ *Magnificently marked wild–caught Royal Blue Discus are becoming rare. This specimen is completely marked and – once acclimated – will become the focal point in an aquarium.*

In the Amazon region this heavy rainfall season starts, as a rule, in December. The rivers reach their highest water levels in January and February, when there is vast flooding in the region. This will last until about June, and water levels start to drop slowly in July and August. The lowest water levels occur during the months of October and November. That, by the way, is also the best time for Amazon expeditions.

Because of the lower water levels the fish are easier to catch at that time. For that reason the main export months for wild–caught discus are from September through March. There are usually no discus exports from May through August.

Clear water rivers have a slightly greenish color and a high degree of transparency. Visibility of up to four meters below the surface is possible. It must also be noted here that the electrical conductivity in these rivers is very low; usually below 15µS. The pH value oscillates between 5 and 6.

Best known to aquarists is the black water of the Rio Negro. Black water rivers have an olive-brown,

← *Barcelos is a sleepy township along the Rio Negro with a population of about 3,000 people. The precise number is dependent upon the fishing season. During the peak season many fishermen are with their boats on the surrounding rivers in order to collect aquarium fishes. Often they are accompanied by the entire family, which is evident from the size and fittings of the boats. The cabin-like superstructures of the boats offer protection against the elements and sleeping accommodations at night.*

↑ The clay-colored water of the flood regions carries lots of sediments. Visibility below the surface is slight. The fish tend to remain close to the surface where they can only be caught at night. Collecting fishes during the day is done by nets. Ideal collecting boats are the shallow-draft canoes, which can take the fish collector everywhere in this region. Navigation with motor boats is often impossible.

→ Small streams and flooded areas make fishing easier than in the rapidly-flowing, major rivers. Blackwater regions with a low pH value are inhabited by few mosquitos, which makes fishing in these areas less difficult. On the other hand, white and clear water regions are infested with masses of mosquitos.

sometimes even dark brown water, which is slightly more transparent than the nearly opaque white water. Visibilities of up to 1 m below the surface may be possible. With 10 to 20 µS, the electrical conductivity of black water is rather low. The pH value is normally surprisingly low; usually it is about 4, but there have also been many pH readings of 3.2 to 3.5. The Heckel Discus especially are often found under these extreme water conditions. This fact must be taken into consideration when these fish are kept in captivity.

During the rainy season it is common for different water types to become mixed, and so suddenly totally new "waters" are being created for the fishes. Factors like this, for instance, tend to trigger spawning behavior in many wild fish species. At that time the fishes of Amazonia can find excellent spawning conditions in vast, flooded areas. In these newly created small and/or shallow bodies of water temperatures will rise due to intense sunlight (sun!). This water temperature increase also stimulates the discus to commence spawning. Due to rising water temperatures there is also enhanced infusorial development, which – together with many other microorganisms – provide a lot of food for newly hatched fish. Since the flooded areas are rarely ever covered by extensive tree growth, water temperatures may rise up to 34°C. This is yet another factor that must be taken into account when breeding fish in an aquarium.

During the main discus exporting months (October through March) many discus specimens are caught in the Amazon region. This is hard work, because seine nets placed in rivers parallel to the river banks must subsequently be freed of driftwood, roots and tree branches. Only after this sort of back-breaking work has been done can the nets be pulled up on

Brown Discus with intense brown coloration. These fish are particularly ideal for planted aquaria, because they provide an esthetically pleasing contrast to the green plants. Discus need not be kept in unplanted tanks.

As seen in this photograph, it is quite possible to keep healthy discus in a planted aquarium. Following a suitable quarantine period and with adequate water quality control there should not be any problems. A prerequisite is also the regular removal (siphoning) of left-over food. Under proper lighting (with red spectral component) the fish will display optimal coloration.

the banks and inspected for fish caught. Many fish collectors go out at night to hunt fish. In fact, it is relatively easy to catch discus and angelfish *(Pterophyllum)* at night. At that time the fish remain immobile close to the banks in shallow water, where they can be spotted in the beam of a flashlight. From my own experience I can confirm that this is an easy way to collect discus.

Collecting discus has become more expensive due to increasing fuel costs and the greater distances fishermen have to travel to catch the fish.

Vast forest tracts have been (and still are) burnt off through prospecting for gold, and this, together with various agricultural and forestry projects, has lead to the massive (and continuing) destruction of discus habitat throughout Amazonia. It is not the fault of aquarists that the wild fish populations are being reduced in the Amazon region; instead, the real killers of nature are the massive forest burn-offs and ambitious tropical timber projects. We all know this, but little is being done about it. Wherever international capital exerts its power there is little chance for nature to survive.

The most important export cities for wild-caught discus continue to be Manaus, Leticia and Belem.

In recent years Santarem has become somewhat of an insider's area for particularly interesting wild-caught discus. Very intensely red-colored discus are caught in the proximity of Santarem. These are the so-called Brown Discus, which the trade refers to as Red Discus because of their shades of brownish-red.

It goes without saying that nowadays large numbers of wild-caught discus are still being exported to Europe. Large consignments also reach the United States and especially Asia. The principal import country for these fish in Asia is Japan. The Japanese pay high prices for exclusive and unusually colored, wild-caught discus. That is also the reason why many discus exporters have concentrated on exporting these fish to Japan. Consequently, top quality wild-caught discus are sometimes no longer available in Europe. The "boom" of wild-caught discus imports into Germany is certainly over. Yet, large numbers of quality-colored, wild-caught discus could still be sold, if these were available. In recent years the interest in wild-caught discus has intensified again, supported by many articles and other publications on these fishes. Maybe aquarists have become tired of brilliantly colored turquoise discus and now would like to see nicely colored, wild-caught fish again.

In those areas where discus occur naturally, they are collected by native fishermen and then kept in plastic tubs. During such a collecting trip the water is changed regularly, so that the fish easily survive the transport for several days.

Once back at home base, the fisherman transfers his catch to net holding pens set up close to the river bank. This way newly caught discus can be kept without problems for long periods of time. Sooner or later a "discus buyer" comes along and picks up these fish and transports them by motor launch to the nearest larger city, where wholesale exporters buy up these wild-caught fish. This exporter will then keep the fish for a while in large concrete vats. Up to that point in time the discus are still in good condition. For shipment overseas large specimens are packed individually in plastic bags. Unfortunately, due to transport cost cutting,

↓ *This imposing Brown Discus shows a large component of blue in its head region and fin margins. The pelvic fins are also intensely red, features that make this specimen particularly interesting. The small red eye indicates that this fish is well conditioned and is not too old.*

→ *The specimen on the right is also wild-caught. These sort of intensely green–colored wild-caught discus were used in the early days to produce the evenly colored, turquoise-colored discus. Jack Wattley was highly successful with his specific, selective breeding program.*

fish are nowadays packed more and more tightly, which often leads to substantial mortalities in transit. Moreover, many fish will be injured or are severely stressed by long transit times and rapidly deteriorating water quality. Discus importers should make genuine attempts to motivate exporters to provide better packing and in-transit conditions for their fish. Once the fish reach the wholesaler, they are placed in holding tanks.

Responsible wholesalers will always attempt to provide newly arrived discus with optimum tank and water conditions. This includes water temperatures of at least 28°C and well-conditioned discus water. Often newly arrived discus are treated with antibiotics as a prophylactic measure, but this is not advisable. After all, why should fish be loaded up with antibiotics if there is nothing wrong with them? It makes more sense to monitor newly arrived fish closely and – if required – initiate a specific disease treatment. Generally, however, this is only ever done at the home of the final buyer, the aquarist. Therefore, aquarists are advised to place newly arrived wild-caught discus first in a quarantine tank where the fish can be observed for some time and possibly be treated, should this become necessary.

Wild discus are accustomed to hiding in dense vegetation. Newly acquired wild-caught discus should be afforded natural or plastic plants in their tanks until they acclimate. Photo courtesy of Living World.

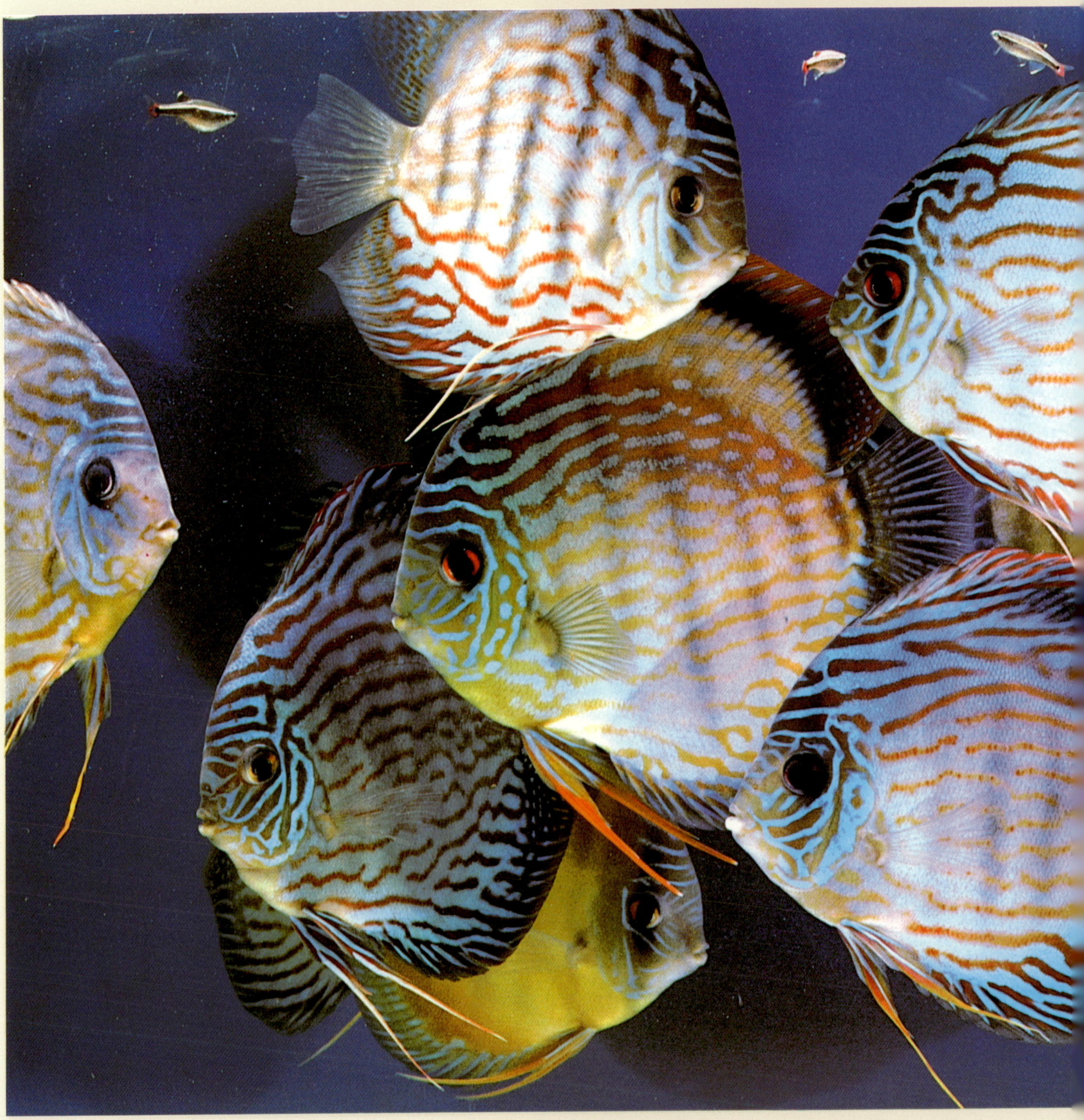

Discus will naturally interbreed, which can be seen in this photograph. All varieties are present, from the common Brown Discus to the conspicuous, red-striped Red Turquoise Discus. This fact leads to the conclusion that the genus Symphysodon embraces only two species, S. discus and S. aequifasciatus.

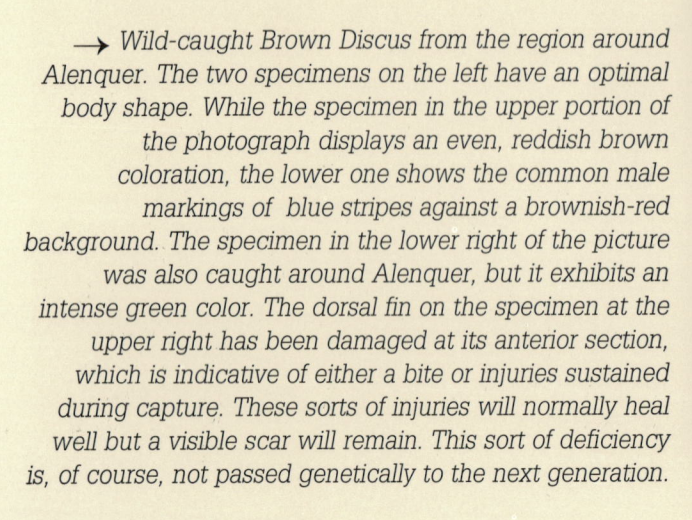

Green Discus with a faint red dotted pattern, from the Tefe region.

→ Wild-caught Brown Discus from the region around Alenquer. The two specimens on the left have an optimal body shape. While the specimen in the upper portion of the photograph displays an even, reddish brown coloration, the lower one shows the common male markings of blue stripes against a brownish-red background. The specimen in the lower right of the picture was also caught around Alenquer, but it exhibits an intense green color. The dorsal fin on the specimen at the upper right has been damaged at its anterior section, which is indicative of either a bite or injuries sustained during capture. These sorts of injuries will normally heal well but a visible scar will remain. This sort of deficiency is, of course, not passed genetically to the next generation.

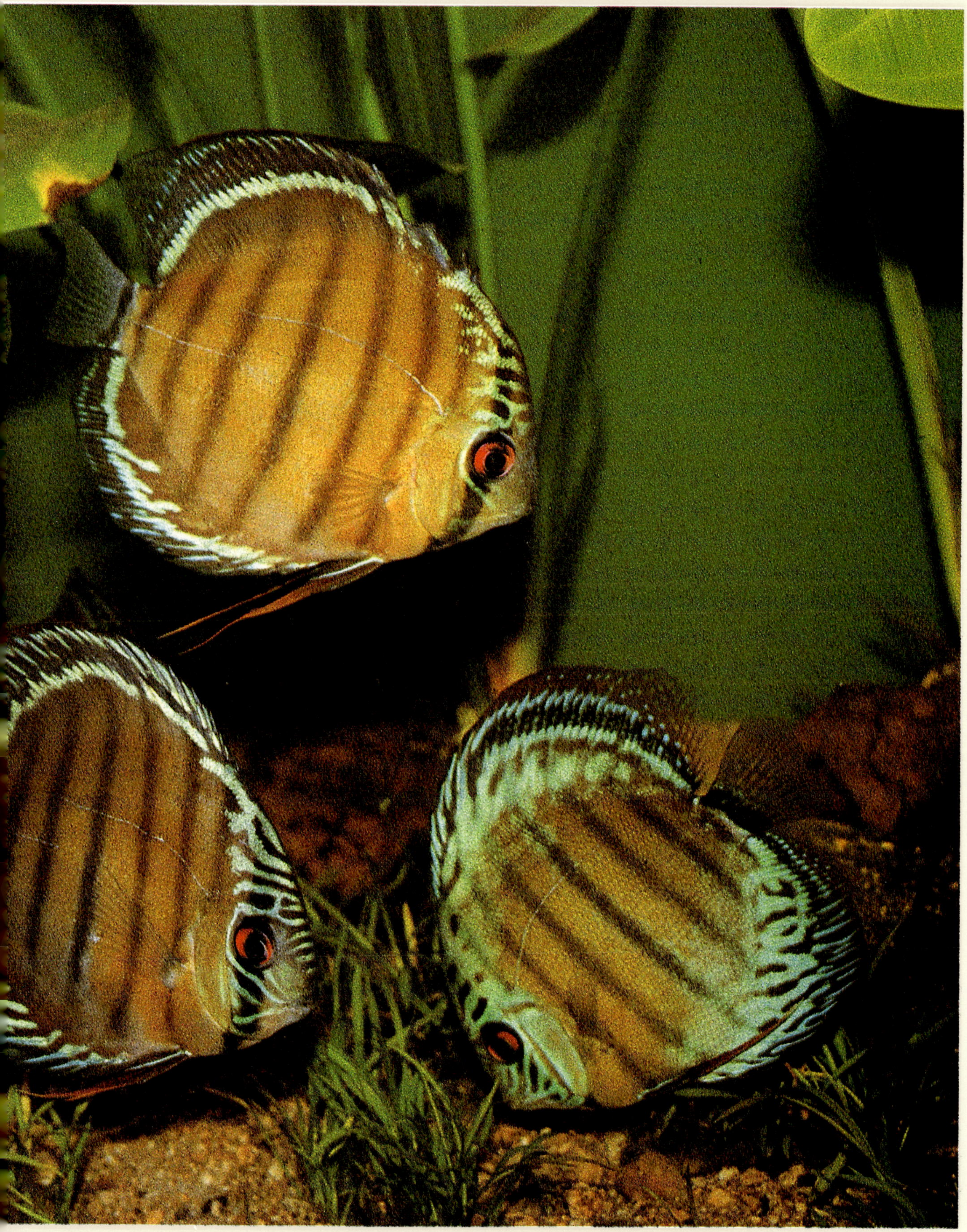

Captive-bred variant from a cross between a Green and a Brown Discus.

Selecting wild-caught discus

If you want to purchase quality discus with good coloration, there are a few things to keep in mind. In terms of coloration, wild-caught discus are significantly different than captive-bred specimens. The latter are generally colored far more intensely. Through numerous in-crossings within the color varieties of turquoise, the turquoise coloration has become strongly dominant in captive-bred stock. In wild-caught discus such intensive turquoise coloration is only ever found in the margins of the fins, if at all.

Wild-caught discus usually do not look well when first held in captivity, because they are generally not kept under optimum conditions. This is reflected in the lower color intensity displayed. That means that you as a buyer will not see optimally colored wild-caught specimens. But do not be misled by these circumstances. Sometimes it may take months of optimum aquarium maintenance for wild-caught discus to achieve maximum color and marking intensity. In fact, it can be be said that these fish will get more attractive from week to week. Consequently, color is not necessarily the first criterium you look for when selecting your wild-caught discus.

The main selection criteria are health and external appearance. Only a close examination will reveal whether a wild-caught discus is healthy or not. Therefore, take your time to observe the fish. Healthy fish are always active and clearly interested in what is going on inside and in front of the aquarium. Flight behavior is completely normal and should not discourage you from buying a particular specimen. It is absolutely essential to observe the color of the feces. Should you see white fecal threads you should be cautious. Pay particular attention to the head region of a specimen. Should you notice even the slightest concave patches in the forehead region, caution is advisable. Specimens with caved-in sides along the abdominal cavity are usually infested by intestinal parasites. Of course, these specimens can be effectively treated, but this usually requires intensive care. It is not necessarily a reason for not going ahead with the purchase of it. An important characteristic for the state of health and age of a discus, especially a wild-caught specimen, is its eyes. They must never appear too large relative to the body size. Moreover, eyes must not grossly protrude from the head. If the eyes are mainly dark or even black with very little red, it means this is an older specimen. On that basis compare individual fish with each other, and then—on the basis of the appearance of the eye — determine a possible age. Since we are talking about wild–caught specimens here, there will be nobody around to give us clues about the actual age of such fish. Specimens with button-like, relatively small eyes are certainly the youngest in the group and should be preferred for a purchase.

This discus is clamping its fins, which shows that is not well.

Do not take much notice of body size. Generally, wild-caught discus remain smaller in nature than captive-bred discus. This is certainly understandable, because captive-bred progeny will be pampered through and through and will normally be given far too much food. Discus in the wild can not compete with such food regimen in excess, and so their body size will remain smaller than captive-bred stock. If the wild-caught specimens you purchase are subadults or juveniles, they can also be raised to substantial sizes with an intensive feeding regimen.

Newly arrived, wild-caught discus often have frayed fin edges. These are either transport or bite injuries, which usually heal promptly and without lasting damage. Severe fin bites, however, will heal up but often leave behind some tissue unevenness. The only effect is esthetic and there is, of course, no effect on the genetics of these specimens.

If discus are sick or refuse to feed for a prolonged period of time, their growth will become distinctly inhibited. This can also be the case with wild-caught specimens. Again, such deficiency is not hereditary. On the other hand, discus with a naturally smaller body may well pass this trait on to many or most of their progeny. Injuries sustained while being reared or held in captivity are not inherited by the progeny. But it is possible that later generations in captivity will become distinctly larger than the original brood stock.

An important feature to watch for when selecting discus is the typical razor-back. If any of the specimens show – when viewed from above or head on – clearly caved-in areas along both sides of the dorsal profile (the so-called knife or razor back), they are obviously sick. Specimens severely affected usually can not be saved. Healthy specimens, when viewed head-on, will have well-rounded sides down from the back line. The head shape above the eyes must be distinctly rounded.

If certain specimens are very dark (almost black), caution is also advisable. These fish may be affected by parasites and refuse to feed. Therefore, it is ideal if you could see the fish actually feed (a reputable dealer will always oblige!). Of course, such darkish coloration can also be a sign of having been frightened very recently. If such fish are healthy their normal coloration will return promptly within a brief period of time.

Healthy discus have dark to almost black feces. Whitish, almost transparent, gelatine-like fecal threads are indicative of a parasite infestation. Such an afflicted specimen is best avoided, certainly by someone who is just starting out with discus.

Gill parasites (worms, flukes) are also common in discus. Should this be the case such an affected fish will normally only breathe with one gill cover; the second one remains closed. Gill flukes are relatively easy to treat and a gill parasite attack is not particularly difficult to get rid of.

If you adhere to the recommendations given here when buying discus and then also plan to keep the fish in quarantine for a period of time, there is nothing to stop you from buying wild–caught discus.

Purchasing wild-caught discus is somewhat different from acquiring captive-bred stock. As you know, fundamentally there are five different color varieties of wild discus. If you favor the Heckel

In 1980 I had already succeeded in producing such red spotted progeny from wild-caught Brown Discus. Not only is the red spotting striking, but the intensely red colored ventral fins and the fin edges are also. It is very difficult to keep these red color characters over several generations. Only by rigid selective breeding is this possible.

Green Discus from Lago Tefe. The even green coloration in the upper and lower region is conspicuous. The only remaining brown is in the anterior part of the body. Breeding stock of this variant can be used to produce even-colored discus.

Discus, it is advisable to stick to Heckel Discus only. This fish, however, places specific demands on the tank water. Since it comes from typical black water rivers it needs water with a relatively low pH value; values from 4.0 to 5.0 are recommended. On the other hand, the electrical conductivity need not be so extremely low as in the native habitat of this fish. A value from 100 to 200μS is totally acceptable. Moreover, water with a higher conductivity remains more stable as far as the pH value is concerned. Experience has also shown that Heckel Discus are particular as far as their diet is concerned. It takes a lot longer to get Heckel Discus to accept substitute foods than it does other wild-caught varieties. Heckel Discus prefer live foods. While this applies to all wild-caught discus, it is even more strongly developed in Heckel Discus. But this does not mean that Heckel Discus will not accept substitute foods such as meat, pellets or flakes at some later stage.

The other three types of wild-caught discus can easily be kept together, since their water quality requirements are largely identical.

All types of discus will interbreed; this is an important point to remember when keeping wild-caught discus. If you want to breed wild-caught discus at some stage later on and wish to maintain pedigree color varieties, you should take that into consideration when you buy your brood stock. Unfortunately, the availability of wild-caught discus fluctuates somewhat. In essence, that means that brown discus, for instance, are sometimes simply not in shops, or there may just

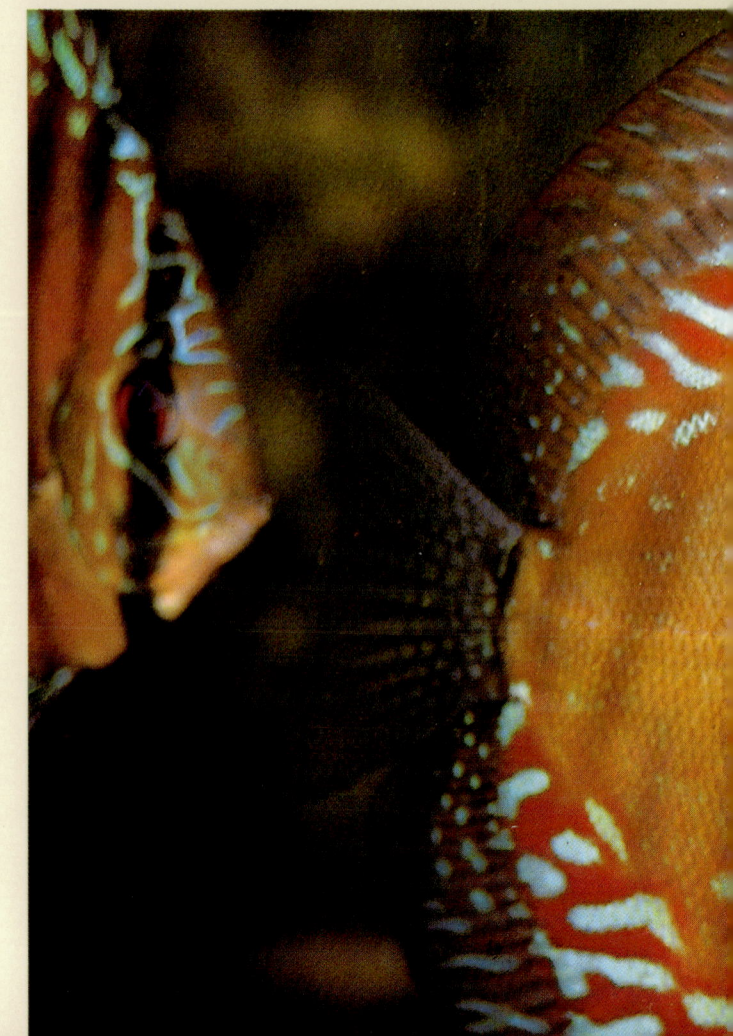

Presumedly a variant cross between Red Turquoise and Green Discus. The body shape is good, but the coloration is unsatisfactory. The middle body section is missing proper markings. ↓

← *Captive-bred progeny of wild-caught discus from Alenquer (from the facilities of Dr. Schmidt-Focke). The deep body shape of these juveniles is conspicuous. In time the red coloration will further intensify.*

Wild-caught discus can easily be kept together with captive-bred stock in the same aquarium. They will mate and breed together. Depicted here are two wild-caught Green specimens from Tefe and two captive-bred, Brilliant Turquoise Discus.

not be any Royal-blue Discus around. Nevertheless, you should always attempt to select the very best specimens from among those fish available and keep them under optimum conditions. You will quickly notice that your wild-caught discus will develop superbly and display magnificent colors. It has been noticed again and again that it takes far longer for wild-caught discus to commence breeding. Consequently, waiting periods of up to two years can be considered quite normal. So, do not get discouraged as far as breeding these fish is concerned. It is simply a question of time and proper care.

Should you have the opportunity to purchase small wild-caught discus, take advantage of it. Of course, such small specimens are not yet particularly attractive, but do not be distracted by that. Once these fish have reached an age of about 18 months they will have developed full coloration and you will not be disappointed. For instance, you could purchase half a dozen small Green Discus, and you would not exactly be envied by your aquarist friends, who would probably not even notice these fish. After one year of optimum care you will suddenly notice that six top-quality discus are swimming in your tank, much to the envy of every discus fancier.

Purchase and care of wild-caught discus requires a certain special intuitiveness or simply a good eye for the development potential in certain specimens.

Quarantine treatment

After new discus have been bought it is very important to keep them in a separate quarantine tank for some time under close observation and – if need be – for treatment. If you buy several wild-caught specimens and keep them together in the same tank without other fish species being present, you can do without the separate quarantine tank. But if the newly acquired discus are going to be placed in a planted aquarium, it is advisable to give them a prior quarantine period. Consequently, the ultimate home for these fish will be largely disease-free.

A quarantine tank should be as large as possible, because the wild-caught discus need to be kept under close observation for several weeks, maybe even for several months. That is particularly applicable if the new arrivals are eventually to be kept together with other discus. Ideally a quarantine tank for practical purposes does not have any bottom substrate. Yet, the fish should be given some hidden places in the form of roots or larger rock slabs. Also providing spawning containers, even ordinary (clay) flower pots, helps to structure the tank and so provides retreats and sight barriers between specimens, which then keeps aggression in check.

Dr. Schmidt-Focke received some Brown Discus specimens with a more strongly defined (fifth) vertical band. The origin of these fish is unknown and so far there are no records of any captive breeding.

Filtration for a quarantine tank should be selected in such a way that there is easy service access and the filter medium can be replaced without difficulties. This seems logical when you consider that you may need to use activated charcoal in order to remove residual medication from aquarium water. Of course, during any disease treatment there must not be any activated charcoal in the filter, otherwise the effectiveness of the medication would be severely reduced. When using medication and chemicals in quarantine tanks, biological activity in the water will be greatly decimated or totally destroyed. This then also destabilizes the biological equilibrium rather quickly. The use of medication also leads to a deterioration of water quality. Fecal matter and the fishes' left-over food must be siphoned out regularly. Yet, while medication is being used water changes must be kept to a minimum. Similarly, large biological filtration units should not be used on a quarantine tank, since the microorganisms in such a filter will be quickly killed off through the use of medications, especially antibiotics. Decay bacteria will reproduce rather rapidly in such "dead" filters. Therefore, it is advisable to use rapid and compact (power) filters.

Aquarists also have a tendency to add several medications simultaneously to the aquarium water. This can have severe consequences, because one does not know exactly what the final combination (of medications) could be. Therefore, it is advisable to apply only one medication at a time, which is then removed again after a few days by filtering the aquarium water over activated charcoal. This should be followed by a partial water change before another medication is added.

Apart from the use of different medications in discus husbandry, temperature increases have also proven to be very effective for treating diseases. Over a period of about 24 hours the water temperature is gradually increased to a maximum of 35°C. This temperature increase places the fish in a fever-like condition, since their body temperature will adjust to that of the surrounding water. Generally, discus will not suffer any ill-effects from an elevated water temperature. Only from a water temperature of 36°C on upwards and with older specimens kept under less than marginal water conditions can there possibly be complications. The fish are kept at 35 °C for two days, and subsequently the temperature is reduced to normal levels again. The ACCURATE measurement of the temperature is very important for this temperature treatment. Rely only on readings obtained with quality thermometers. During such heat treatment it is important to make sure that only small amounts of food be given; left over food will decay rapidly! Simultaneous treatment with medication during the heat treatment is not advisable. Please, also make sure that there are no remnant medications (from previous treatments) left in the water before heat treatment commences. A partial water change (at least 1/4 of the tank volume) is recommended at the end of the heat treatment.

It is also advisable to filter the tap water over activated charcoal prior to using it in a (partial) water change. This procedure has proven to be very effective, because nowadays tap water is so overloaded with various contaminants that many fish will react adversely or display allergies to it.

After you have accommodated your newly acquired wild-caught discus in a quarantine tank, it is advisable to wait for a few days, closely watching the fish. If they settle in properly they will also start

Newly imported wild-caught Blue Discus, kept in a quarantine tank for observation. A neutral bottom substrate was added to the tank to reduce flight behavior. A sufficiently long quarantine period is particularly important for wild-caught discus.

feeding promptly. Initially their favorite food will be chironomid larvae. Although these larvae can be contaminated by environmental pollutants, it is still recommended to feed them to newly arrived wild-caught discus, at least at the onset of feeding in captivity. The feeding stimulus created by this type of fish food is usually (for the fish) so overpowering that even those specimens reluctant to feed can be persuaded to take this food. You may also try glass worms and mosquito larvae. Eventually the discus will take just about any type of food, but at the beginning one may need to be patient with these fish. Also, keep monitoring the fecal discharge. As long as it is dark, virtually blackish, there is little indication of an intestinal parasite infestation. However, if watery, white fecal threads appear, the fish must be treated. Intestinal parasite infestations can be treated effectively with two different types of medication; however, they are only available on prescription (contact a veterinarian). You should first try to bathe the fish for a prolonged period in Metronidazol, commonly available from chemist shops and drug stores. It is a white powder which is dissolved in lukewarm water and is then added to the aquarium, at a dose rate of 250 mg Metronidazol per 50 liters of tank water. This medication should be left in the tank four to five days, and then followed up with a partial water change.

The second medication against intestinal parasites is Flubenol (5%). This is also a powder, which—after it has been dissolved in a small amount of warm water—is added directly to the aquarium water, at a dose rate of 200 mg per 100 liters of aquarium water. It must be noted here that Flubenol (5%) contains only five percent of the active ingredient flubendazol. That means from a total weight of 200 mg medication only 10% flubendazol is actually added to 100 liters of tank water. Flubenol (5%) must be applied three times in a row. The first application of this medication should remain in the tank for seven days. This is followed by a partial water change, and then the second treatment is applied. Again, after seven days part of the water is changed and the third treatment is added, also for seven days. The entire treatment period extends over 21 days, which is necessary to kill the eggs of the parasites as well as any gill flukes. After the treatment there has to be another partial water change, and the entire tank volume should then be filtered over activated charcoal.

Should your wild-caught discus have incurred obvious skin injuries, these are simply treated with a combination of malachite green and trypaflavine. Most drug stores or chemist shops will mix you a stock solution of 5 g trypaflavine and 50 mg malachite green oxalate, dissolved in 1000 ml of water. This is sufficient for 1000 liters of aquarium water. This solution is then used at a dose rate of 1 ml per 1 liter of aquarium water. The active ingredients in this stock solution can be absorbed via human skin. Consequently, it is important to wear rubber gloves when handling these chemicals (including the stock solution). This medication is left in the tank for three days. Then make a partial water change and remnants of the chemicals are filtered out afterwards via activated charcoal. Please note that this medication will also damage plants. It is also very effective against skin flagellates, which are killed within the recommended three–day treatment period.

A well-conditioned aquarium also carries populations of useful and generally harmless microorganisms. Best known among these are the planaria, which sometimes occur in plague proportions. They are usually transferred to an aquarium via fresh–frozen foods, and are difficult to get rid of. Fortunately, planaria occur only periodically. But they are more common in tanks with a large fish population and correspondingly large food regimen. The microbial world of aquarium water also includes such harmless protozoans as bell-shaped *Vorticella*, rotifers and heliozoans with their radiating, sun-like appearance; although when viewed under a microscope these organisms may appear to be anything but harmless. Consequently, you should not be misled when you look through a microscope at a drop of water from your discus tank, and don't start adding new medications every time you see an unusual organism!

After a successful quarantine period of a few weeks the healthy discus are transferred to the regular tank, provided of course the fish will accept all standard discus foods.

The origin of this Brown Discus is also unknown. Conspicuous for this specimen is the fact that the black band in the fin margins is far more strongly defined.

Juvenile, wild-caught Green Discus.

Setting up a Discus Tank

One of the first things to be resolved before starting to set up a discus tank is the question of whether this is going to be a display tank or simply a holding tank. The easiest way of keeping and eventually breeding discus is in a tank totally devoid of any bottom substrate. If we also want to add plants to such a tank, they can be planted in shallow flower dishes which are then placed directly on the glass bottom of the tank. With the aid of roots and spawning pots we can delineate territories for the fish. The size of the tank is dependent upon the number of discus to be kept. Ideally, we should offer a volume of 75 liters for an adult discus, and it goes without saying that a discus tank must never be too small. Experience has shown that a tank with the minimum dimensions of 1.2 m long x 0.5 m wide x 0.5 m high is essential for adequate discus husbandry, and larger dimensions are better yet. In fact, the ideal discus aquarium is the 2–meter-tank, with the dimensions of 2 m long x 0.6 m wide x 0.6 m high. In such an aquarium we can easily keep eight to ten wild–caught discus without any problems.

The second variation of a discus tank is the planted aquarium. By combining the fresh green of water plants with the dark wood color of submerged roots we achieve a magnificent esthetic picture, where the discus provide a distinct color contrast.

Securing plants on the bottom requires the introduction of a layer of sand, of at least 4 to 6 cm thickness. Since tropical water plants do not like to get "cold feet" it is advisable to install a thermostatically-controlled heating cable in the substrate. Fine sand has the advantage that leftover food can not penetrate into the substrate and decay cannot take place. For esthetic reasons you should select for the foreground (and indeed for the largest part of the tank) small-growing plants which form a virtual "lawn" over most of the bottom. Along the back of the tank we then position a few selected taller plants as solitary optical focal points. Do not offer too many hiding places for the fish among the plants, because they will only be too willing to seek them out and hide in them, and you will see less of your fish!

For water plants to grow properly it is important to plan for supplementary carbon dioxide to be added to the tank. Select a large-capacity unit which will be more effective in achieving luxurious plant growth. Many different sizes and models are available and your aquarium fish dealer can advise you. Discus can handle carbon dioxide fertilization without any problems. Since (most) water plants prefer low carbonate levels in the water, a discus tank can easily have a carbonate hardness of up to 3° German hardness. The pH values will settle down somewhere between 6.0 and 7.0. Such a planted aquarium must also be given regular partial water changes, ideally about once a week with a minimum volume of about 10% of the entire aquarium content. Better yet is a larger partial water change of 20 to 30%. When doing this you will notice that there is better plant growth and your fish seem to appreciate it too. It goes without saying that regular water changes are also important for discus tanks without water plants.

There are no special lighting requirements for the discus tank. Fluorescent tubes seem to be the preferred light source. It is totally wrong to keep discus in tanks with diffuse lighting. Under such illumination the fish have a tendency to become shy and hide in corners. The more activities there are around the tank, the less shy discus become.

FILTRATION

It need not be stressed that an aquarium must have proper filtration. Numerous systems are available from aquarium fish and pet shops. Aquarists specializing in planted aquaria generally recommend keeping the filter small, together with more frequent partial water changes. In Asia most discus aquaria do not have any filtration at all; instead there are daily, substantial water changes. At least 50%, but usually more like 80% of the aquarium water is replaced. Since we can not adopt such a system for (obvious) economic reasons, it is important to plan for a high-capacity filtration system for our discus tank.

← Chemi-mat is a unique product which removes excess phosphates from the aquarium. Photo courtesy of Boyd Enterprises.

Cycle is a bacterial solution which attacks aquarium sludge through natural, bacterial action. It should be added weekly to the aquarium water. Photo by Hagen. →

It is usually necessary to change the water in a discus aquarium on a regular basis. This is easily accomplished with a NO-SPILL aquarium siphon. Photo courtesy of Python. ↓

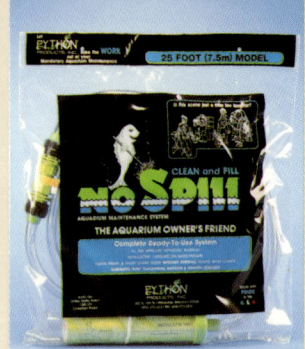

California Aquarium Supply Co. makes a deluxe aquarium kit perfect as a discus aquarium. ↓

Vita-chem is a superior brand of vitamins to assist in acclimating wild-caught discus, as well as keeping them healthy. Photo courtesy of Boyd Enterprises. ↓

← Cleanliness is mandatory for discus tanks. Cycle is a filter supplement with organic means of removing sludge. Photo courtesy of Hagen.

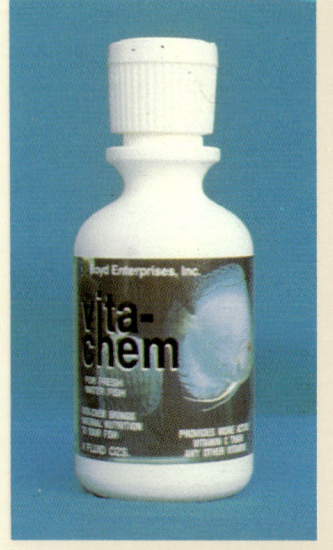

The large rectangular aquaria made by California Aquarium Supply are suitable for discus since discus need lots of room. ↓

Discus require clean water under constant filtration. The AQUAMASTER 250 by Danner, is ideal as an outside filter for discus tanks. Photo courtesy of Eugene Danner. →

← *FIN-CARE, which is a water conditioner, should be used with all fresh water before the discus are placed into it. Photo courtesey of Hagen.*

To supplement or replace an outside filter use the canister filter. Supreme makes a range of these filters with a size to fit every discus set-up. Photo courtesy of Eugene Danner. →

Discus breeders with several tanks tend to favor filtration systems with multiple chambers and a prefilter. Water from the tank flows into the prefilter chamber via an overflow. The medium here is usually filter wool, which can easily be washed or replaced on a daily basis, if need be. From the prefilter chamber the water flows through several filter chambers, each with a different filter medium. Every aquarist seems to have his or her own recipe here. For instance, the filter medium for the filter chambers can be clay pipelets, sintered glass, plastic filter medium bodies, lava split or peat moss. The water, which has passed through all the filter chambers, is then collected in a clear water chamber (or reservoir), and is from there pumped back into the aquarium. If the prefilter works properly, very few dirt particles and debris reach the subsequent filter chambers, and so the medium in these chambers need not be attended to for long periods of time. In fact, usually they can be operated without problems for a year or so. These sort of filters are often referred to as biological filters. The designation "biological filter" is indicative of the fact that in such a filter organic waste products are being decomposed (oxidized) by microorganisms. Nitrifying bacteria will also become established in these filters. Unfortunately, these biological filters do not work well at low pH values.

Biological slow-filters for breeding tanks can also be a simple foam cartridge filter. A foam cartridge is pushed over a plastic pipe and is driven by air supplied by an air pump. The rising air pulls water along with it and forces it through the foam cartridge. Nitrifying bacteria will become established in the tiny pores of the foam rubber. This sort of filter functions well provided there is a constant flow–through of oxygen-rich water. However, after a few weeks the foam rubber cartridge will become clogged and must be washed out. This must be done in lukewarm water so that the bacteria are not washed out or damaged. These foam cartridge filters are particularly useful for breeding pairs with young.

There is also the possibility of using rapid filters, which will clear up the water mechanically and provide for an excellent optical appearance of the

California Aquarium Supply Co. makes a Biological Wet Dry Trickle Filter which is perfect for keeping the water in a discus tank clear. BIO-SPHERES by Hagen are plastic balls upon which bacteria grow when the spheres are used in a filter. These bacteria aid in keeping the water clean and pure since they eat the debris found in an aquarium.

water (without, however, improving chemical water quality).

Biological inside filters, driven by a pump, are now being used on an increasing scale. The aquarium water is led through various filter chambers and then pumped back into the tank.

The smaller the filter is, the more frequent water changes have to be made!

THE CORRECT WATER

For normal aquarists using tap water does not pose any problems. Drinking water from municipal water supplies is generally adequate for use in aquaria. In fact, captive-bred discus can be kept in soft to medium hard tap water. But if you want to keep – or even breed – the more delicate wild-caught discus, the tank water must be properly conditioned. Some tap water supplies are strongly chlorinated. If this is the case, the chlorine must be removed before the water is used for such a discus tank. This is easily done by filtering the chlorinated water over activated charcoal. In fact, it is advisable to filter all tap water to be used in discus aquaria first over activated charcoal.

For that purpose you can easily adapt a filter canister or plastic cylinder, by filling it with quality activated charcoal. The tap water is then simply run through the charcoal. This also removes other harmful substances from the water. Once the chlorine has been removed, the water can be used for the discus tank, provided it is of appropriate hardness. If it is too hard, the hardness level will need to be lowered.

If the hardness of the water is below 10 degrees total hardness, partial softening of the water can be

The so-called peat bomb. The discus water to be conditioned is forced slowly through a container with peat moss. This causes a slight reduction in water hardness and the water is enriched with residual peat moss material. The pH value may also drop slightly. Many breeders, including Dr. Schmidt-Focke, tend to swear by this method. It need not be stressed that the peat moss used must not include any fertilizers or other additives.

achieved by filtering it over peat moss. The peat moss filter is the simplest procedure for softening water. Best suited for water conditioning is the light brown, white peat moss from higher elevations (high moors). It removes the hardness from the water and adds humic (ulmic) acids. Consequently, the pH value will drop slightly. It is, however, of paramount importance that you make sure that the peat moss you are using does NOT contain any fertilizers or similar additives. For use with a small tank, the peat moss can be placed directly in the filter chamber. If you wish to use peat moss externally, you can place it inside a commercially available canister filter (instead of some other medium) and then simply filter the entire tank volume through this filter. Many discus breeders use a so-called peat bomb. Peat moss is placed inside a larger container, and the water is forced (under pressure) through this peat bomb. If you own a large water barrel you can simply use that to bring peat moss into contact with water. When using this method, it is advisable to suspend a small air stone in the barrel so that the water is constantly kept in motion. Subsequently, you can drain the peat moss treated water via its drain valve or simply siphon the water out and transfer it to the aquarium. In order to avoid excessive stirring up of the peat moss in the barrel it is advisable to place the moss into nylon stockings, securing it with a knot at the upper end of the stocking. This leaves sufficient contact between water and peat moss, without actually contaminating the water with peat particles.

This technique (using nylon stockings) can also be employed on a smaller scale for breeding tanks or in outside filters. I have frequently used half a liter of peat moss placed in such a stocking and suspended that directly in the breeding tank. By keeping it inside the tank for days or even weeks the peat moss will gradually modify the surrounding water. This method has often created surprising spawning results. For large multiple-chambered outside filters it is quite possible to place several large peat moss bags in some of the chambers and then remove them again as need be. It is, of course, imperative to constantly monitor soft water, because it is less stable and can rapidly produce pH variations.

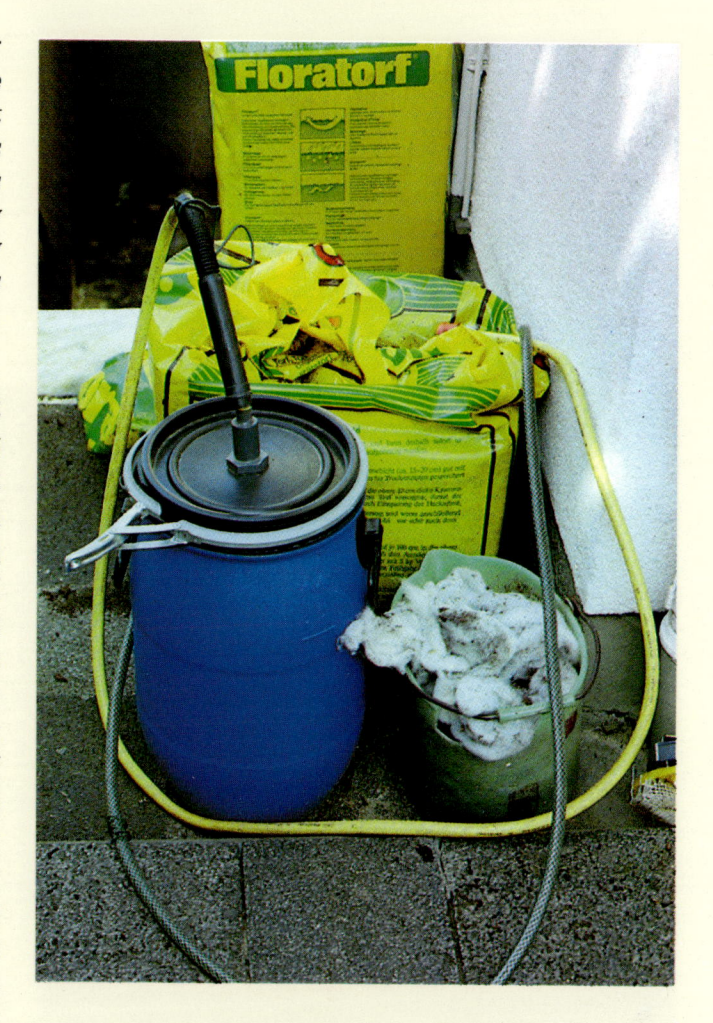

An older method to soften aquarium water is through the use of ion exchange resins. There are two different types of ion exchange resins: cation exchangers and anion exchangers.

A partial desalination is indicated for water with a high carbonate hardness. If the total hardness is made up of at least 80% carbonate, it is advisable to soften the water by means of a cation exchanger. The water flows over exchange resins which affect a constant chemical modification of the aquarium

Wild-caught Green Discus from Tefe with conspicuous brown base coloration. The red dots along the flanks and in the fin margin are still weakly defined. An optimal feeding regimen with carotene-containing food items tends to lead to an intensification of these dots with increasing age.

water. Especially advantageous are resins which contain indicator dyes. As the resin becomes chemically exhausted it changes color from light brown to red. By using a transparent exchange column the remaining (exchange) capacity is easily visible. The flow rate of the water to be treated, through the ion exchange column, must be very slow if it is to be fully effective.

Depending upon whether it is a cation or anion exchange resin one uses an acid or base solution to regenerate it. So, for instance, 1 liter of the resin is regenerated with two liters of 10% hydrochloric acid. This acid must flow slowly through the resin, in 20 minute intervals. Subsequently, the regenerated resin must be rinsed out with about 10 liters of tap water over a period of 30 minutes.

Another, still more weakly acidic cation exchanger has an even slower flow-through rate, but it only needs to be generated with 2 liters of 3% hydrochloric acid. These weakly acidic cation exchangers can even be regenerated with weak acids (e.g. citric acid) instead of hydrochloric acid. For instance, one liter of a weak resin can be regenerated with two liters of 6% citric acid solution. This procedure takes 30 minutes and the rinsing phase with 10 liters of tap water requires one hour.

TOTAL DESALINATION

During this desalination procedure the tap water flows first through a strongly acidic cation exchanger, which is loaded with hydrogen ions. This exchanger replaces all cations with hydrogen ions and so converts all salts into their respective acids. This acidic water is then passed through a strongly basic anion exchanger, where all anions are replaced with hydroxide (OH) ions. Of course, both ion exchangers must be operated (and regenerated) independently of each other. Subsequently, it must be rinsed with at least 15 liters of water over a one-hour period. It is important here that desalinated water is used for the rinsing step. Full compliance with manufacturer's instructions and recommendations is essential.

Full desalination is a complex exchange procedure, because it requires two completely separate exchange resins. Acquisition and maintenance costs are high, but full desalination provides nearly salt-free water, which has an extremely low conductivity, usually below 5µS. The advantage of full desalination over partial desalination lies in the fact that the raw water can be mixed as required. Using fully desalinated water (together with other types of water) you can mix your own, individual aquarium water. It must be noted here that for osmotic reasons fully desalinated water can not support any life forms, and must be "cut" with regular water. For instance, spring water is a good source as a mixer with fully desalinated water.

REVERSE OSMOSIS

In recent years a rather interesting water conditioning technique has been used increasingly by aquarists, that of reverse osmosis. Actually this is a technology that has already been around for a long period of time, but became popular with aquarists recently as the price of equipment came down to more realistic levels for a reasonably powerful unit. Of course, these reverse osmosis units cannot supply huge volumes of aquarium water, but they are adequate for small breeding facilities. Professional breeders, however, will need to invest in a large reverse osmosis unit some day, if they wish to stay competitive with other breeders.

Desalination units are important for many discus fanciers in order to condition the water properly. The unit at the right contains the indicator resin. The color change in the exhausted (red) resin is clearly visible. Regenerating the resin must be done with considerable caution. Freshly regenerated resin must be rinsed out thoroughly prior to re-use.

The functional centerpiece of a reverse osmosis unit is a semipermeable membrane. Tap water contains dissolved salts and other solubles, which can not penetrate the fine membrane. By means of high pressure the water is pushed through this membrane, resulting in pure water on one side and a higher concentration of dissolved salts on the other. Over-simplified, one needs to imagine this procedure as being some sort of ultra-fine filtration.

Since reverse osmosis units for aquarium purposes work well with normal water line pressure, it is normally not necessary to install a pump for additional water pressure. But, of course, by using a pump the water pressure would be increased, which then enhances the effectiveness of the unit.

The membrane material is usually a polysulfonated-polyamide, which has an excellent separation capability, but it is sensitive to chlorine. Therefore, it is advisable to install an activated charcoal filter on the water intake side of the reverse osmosis unit. Also recommended is the installation of a pre-filter for the removal of suspended particles on the intake side of the activated charcoal filter.

Small reverse osmosis units for use with aquaria have three pipe connections:

1. Intake for tap water.

2. Discharge of desalinated (pure) water (also known as permeate).

3. Discharge of salt-containing effluent (also known as concentrate).

The pure water still contains less than 10% of the dissolved substances. It is similar to distilled water and is also largely free of pesticide remnants. Unless reverse osmosis units are operated continuously, the water must be kept inside the unit while not in operation. If the membrane is permitted to dry out it may sustain damage.

Small reverse osmosis units, which operate with normal line pressure, return about 20% of the used water as permeate. The remaining 80% is going to waste or is utilized for other purposes, i.e. for watering the garden.

If a reverse osmosis unit is being protected by an appropriate pre-filter, it has a longevity from about three to seven years. Through the installation of technical accessories the capacities of these small units can be increased, so that they can meet the increasing demands from aquarists specializing in discus.

Reverse osmosis removes up to 90% of all dissolved matter in water, which lowers the pH value only marginally. Nitrate, an undesirable compound in aquarium water quality, is repressed. Desired trace elements or iron can be subsequently added again to aquarium water, as specifically required. Pure reverse osmosis water must not be used undiluted, especially not for fishes from soft water regions. Consequently, small volumes of tap water or spring water are mixed with reverse osmosis water. This enables you to mix your very own, individual aquarium water. In essence, it is important to know and understand that discus are easy to breed in such soft water.

The AQUAFIN Reverse Osmosis Water Purification System is a necessity for the serious keeper of discus. →

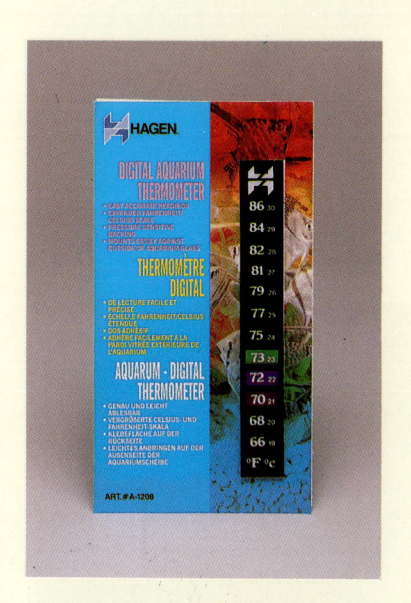

↑ Discus require higher than normal aquarium temperatures. Temperature fluctuations are especially bad. Hagen Digital Aquarium thermometers allow you to accurately monitor the temperature of the water in a discus aquarium.

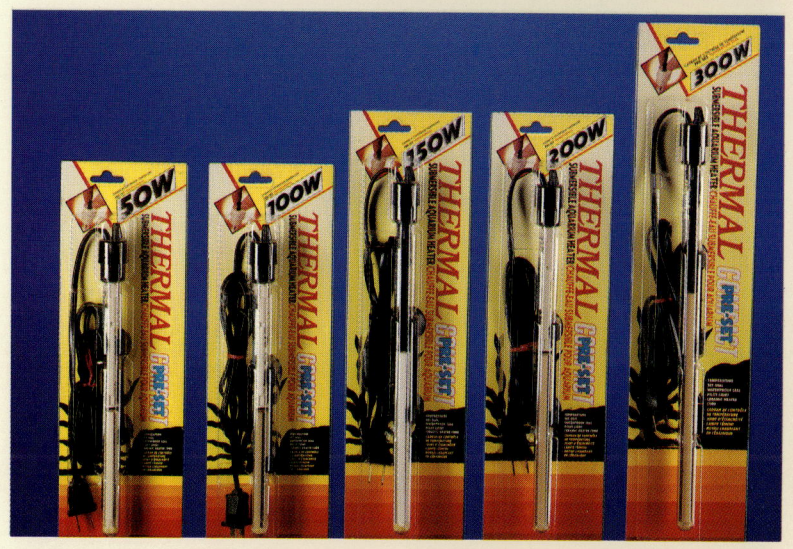

Hagen makes a heater in various wattage for discus tanks of any size. These are submersible, thermostatically controlled heaters which can be strategically placed in the discus tank. ↑

Energy Savers manufactures a complete range of fluorescent lamps which allows you to view your discus under ideal conditions. →

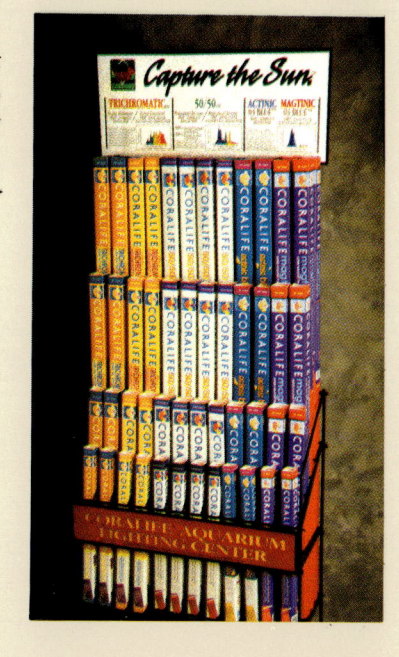

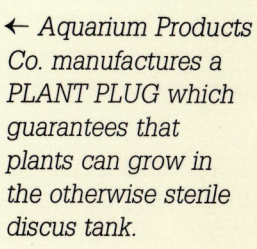

← Aquarium Products Co. manufactures a PLANT PLUG which guarantees that plants can grow in the otherwise sterile discus tank.

Care and breeding of wild-caught discus

Wild-caught discus certainly have higher, more specific requirements in terms of their care than captive-bred specimens. That seems to make sense since we are dealing with wild fish, which a few weeks earlier were still at liberty somewhere in the Amazon River system. In order to provide these fish with optimum conditions, the aquarium selected for them must not be too small. Optimal water conditions and adequate filtration are essential prerequisites for the well-being of these fish. When I started out importing large numbers of these magnificent wild-caught discus about 20 years ago, I noticed that these fish seemed to do best in large, well-established tanks. So, please do not make the mistake of putting discus in newly set-up tanks.

During the first few months the most important thing is simply to look after and observe these magnificent fish. Quite some time will pass until they have become acclimatized and have truly settled down. Food and feeding problems must be resolved and – if need be – pathological conditions need to be controlled. It would be far too early at

If several adult wild-caught specimens are kept together, chances are that a naturally-mated pair will develop from such a group. This is an ideal method of pair formation because it virtually guarantees compatibility during breeding. It may take up to two years for wild-caught discus to breed in captivity. ←

↓ Sparring and other temporary disharmonies are completely normal in a discus aquarium and usually end without physical injuries. The fish will swim toward each other, spreading their gill covers and showing threat displays. The intimidated specimen will quickly turn away and make room for the dominant specimen.

that stage to even consider breeding the fish. It has been observed over and over again that breeding wild-caught discus generally is not successful for a year or two. Wild-caught fish simply require more time to adjust to life in captivity and to breed with any degree of success. These fish are still largely dependent upon meteorological influences, temperature variations and seasonal conditions. You must give wild-caught discus time.

If you keep several specimens together in the same tank, there will be hierarchial aggression and possibly some day the formation of a bonded pair. Once such a pair has been formed in a large tank, the two fish will keep to themselves in a corner of the tank and there begin spawning preparations. Daily, close monitoring of the tank and its occupants will quickly reveal and identify a newly formed pair. Sex determination of discus is nearly impossible, and it is harder yet in wild–caught fish than in captive-bred stock. Among the latter one can often use size and finnage as possible sex characteristics provided the fish are of an identical age, but for wild–caught specimens this is really

74

Pair bonding is not always peaceful. The female on the right is putting the substantially larger male under pressure. The female is the dominant partner during spawning preparations. Later on this can change quickly and the male may take over dominance in the aquarium. Ramming by the female is not dangerous to the male, and indeed this can be interpreted as a lover's quarrel. ←

difficult, if not impossible, since we rarely ever know the age of these fish.

Once a pair has been naturally formed, the question arises whether this pair should be transferred to a separate tank. I would not recommend it, because the stress involved would traumatize the fish and destroy their established rhythm. Just leave everything as it is. Should there be too many other discus specimens in the tank, it may be advisable to remove some of them. But leave at least two other specimens in the tank, so that there is a "natural enemy" for the pair.

Clay vases or flower pots have proven to be highly effective as spawning substrates; however, discus will also accept other objects to deposit their eggs on. Stone and plastic tiles or parts of tree roots are all readily used as spawning substrates by these fish.

From then on the water quality parameters must also be closely monitored and should be recorded in detail. Wild-caught discus tend to react to changes in the aquarium water, for instance, by suddenly starting to spawn. In fact, you can possibly stimulate your fish to spawn by slightly varying the temperature, marginally changing the pH value or through water changes. These are some of the phenomena that happen in nature, particularly at the onset of the rainy season where there is a massive influx of fresh (soft) water in the habitat of discus. In fact, wild-caught discus normally spawn in flooded areas in the Amazon region. Even an approaching atmospheric low pressure front tends to favor spawning.

If you are among the fortunate aquarists who were lucky to get a pair of wild-caught discus to spawn, you can be genuinely proud of yourself. This is indeed still one of the most difficult tasks when keeping a tropical freshwater aquarium. You can recognize the early spawning preparations by the color change in the two fish. Both will display a darker coloration especially along the posterior half of the body. The last four vertical bands become more distinct and the tail looks smoky-colored. When this color pattern appears one can anticipate an early spawning. The fish also start to clean a prospective spawning substrate, by frequently approaching a particular site and picking dirt and debris from it. Occasionally there is a distinct trembling of the body. With their fins twitching the two fish approach each other. Everything looks very harmonious. Sometimes it looks like the two fish are nodding to each other. During this phase the female keeps approaching the spawning site and makes some simulated spawning passes without depositing any eggs. Just prior to the onset of spawning a large spawning tube appears at the anal region of the female. This wide tube protrudes about three to four millimeters from the body. The fertilization tube of the male is much shorter and more pointed. The female continues making simulated spawning runs, apparently to entice the male to follow. If the spawning activities are being disturbed by the other discus in the tank, it may be necessary to remove them. It is counter-productive if the male is constantly distracted by the presence of natural enemies and so cannot fertilize the eggs properly.

These spawning preparations can last for a period of several days; however, sometimes things happen very quickly. Discus spawn generally during the evening hours. The female approaches the spawning site from below and on her way past and upward she deposits a string of about a dozen

The genital papilla of the female is larger and more roundish. It protrudes about 3 mm from the body profile. It is only visible during spawning, which makes it impossible to determine the sex earlier. The female approaches the spawning substrate in an almost vertical position (i.e. from below) and the eggs extrude from the spawning papilla directly onto the substrate.

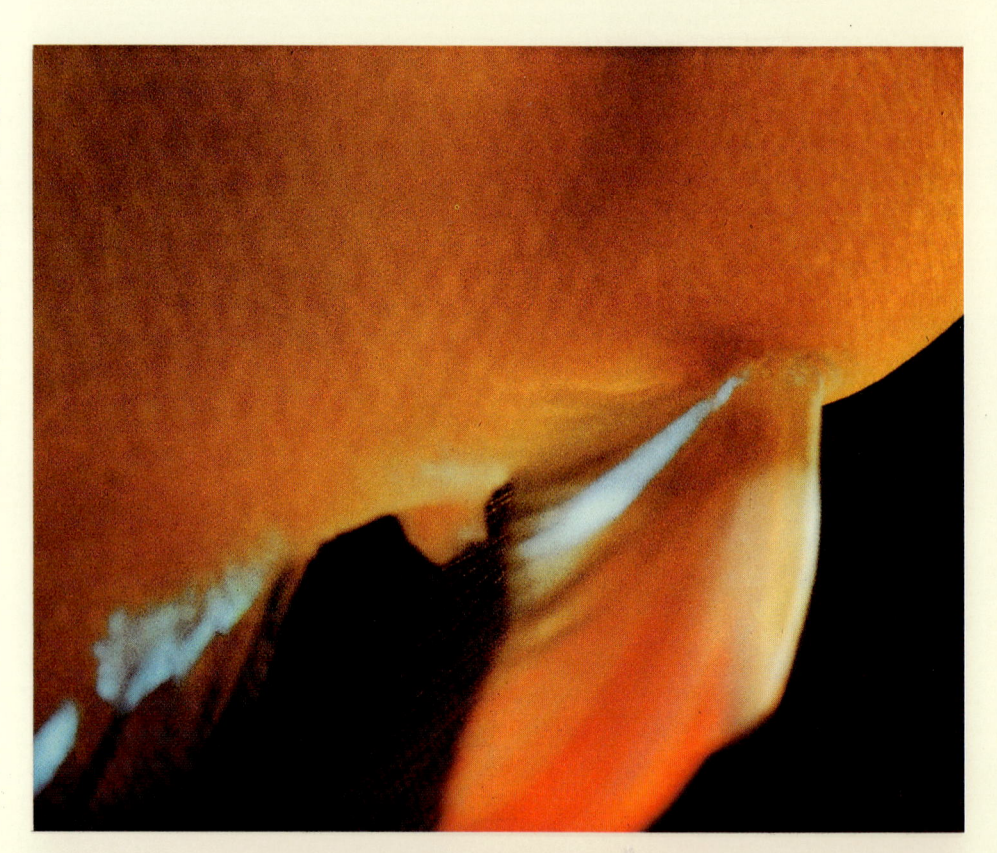

The fertilizing papilla of the male is slightly pointed and smaller than that of the female. It, too, is only visible during actual spawning when the male fertilizes the newly deposited eggs with his sperm.

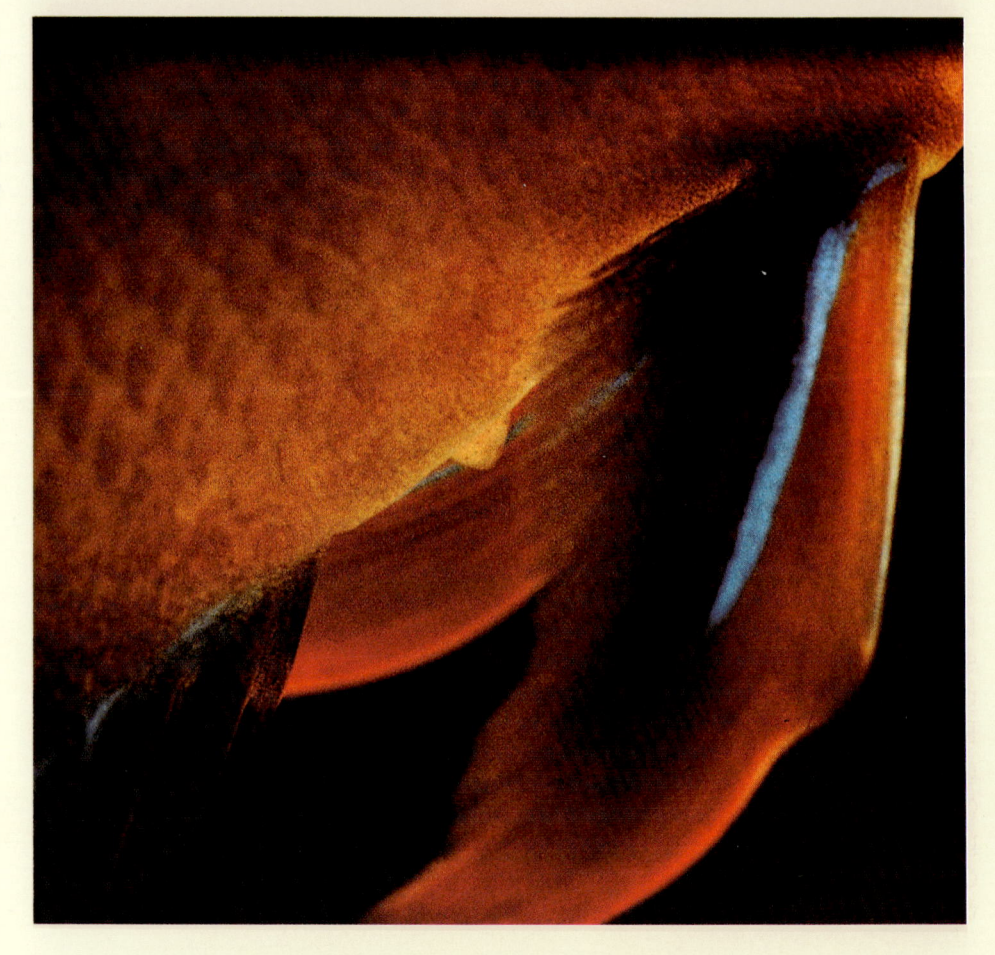

A pair of Alenquer Discus commences spawning.

eggs or so. Once the female has deposited one or two strings of eggs like that, the male should start fertilizing them. A well-matched, harmonious pair will alternate in depositing eggs and fertilizing them. If at all possible, attempt to observe whether the male actually fertilizes the eggs. The average clutch of discus eggs contains about 200 to 300 eggs. During the spawning procedure any strong water current in the tank should be avoided. Turn the water flow down or even turn off the filter for an hour or so.

During the actual spawning you will also have had the opportunity to determine with accuracy the sex of the respective fish. It is also advisable to photograph the fish, if possible, for the record. Further specific characteristics of each can be later recorded and used for comparison purposes. The greater the compatibility between the two partners of a bonded pair the better the survival chances of the young. Wild-caught discus have a very strongly defined parental instinct. Under optimum water conditions such a clutch of eggs will develop normally and the newly hatched larvae will eventually grow into strong young fish.

Once spawning has been completed the adults will hover in front of the clutch, fanning the eggs with their pectoral fins. Both partners will take frequent turns at this. Should there be too much aggression between the parents, one may have to be taken out of the tank. Generally speaking, wild-caught discus are better in parenting than captive-bred stock. The most important prerequisite for successful discus breeding is, however, to have a well-bonded, highly compatible pair and to offer optimum conditions.

At a temperature of 30° C the eggs require about 55 to 60 hours for their normal development. About 24 hours after spawning you can recognize the developing dark eyes in the eggs. Following a development period of 60 hours the larvae will hatch, tail first, and then hang attached from the

After the spawning substrate has been cleaned, the female starts to deposit her eggs on it. The male observes closely what is happening and must no longer be disturbed. Somewhere from 200 to 300 eggs are produced per spawning.

spawning substrate. The tails of the larvae vibrate very strongly; however, with the aid of adhesive glands on the head they manage to remain attached to the spawning substrate. The parents will frequently relocate their young. The young are gathered up by mouth and are gently "chewed" and transported to a new site. But this is normal behavior. It is also advisable to leave a night light on in the room after dark. While providing brood care the adults feed only intermittently and take very small amounts of food; this needs to be taken into consideration when feeding and in the type and amount of food given.

As soon as the young start swimming free, after about another 60 hours, it is important that they swim toward their parents. If this does not occur the young are generally lost. Sometimes this does happen and it invariably contributes to unfavorable water conditions. The parents develop a skin secretion on their body surface which consists in part of bacteria. The young will feed on this secretion, which is their very first food. Without it newly hatched discus cannot normally survive. Rearing young discus artificially should not be a standard goal among discus breeders, but done only in an emergency. The skin secretion is required by the young as their sole food for the first four to six days. The bacteria present in the skin secretion assures that the digestive system of the young fish starts to function properly. Five to six days after the young discus have started swimming freely we supplement their diet with newly hatched *Artemia* (brine shrimp) nauplii. Once the little fish take this supplementary food without difficulty, they can be separated from their parents. It is, however, advisable to keep the young with their parents for another two to four weeks. If the parents are healthy there is no risk of transferring any

↑ *The newly hatched discus larvae are clearly visible. The female intensely guard her brood.*

Wild-caught Alenquer female guarding her brood. The successful breeding of wild-caught discus is a high point for any discus fancier.

Top: Successful spawnings of wild-caught Heckel Discus are rare. They are indeed the most difficult discus to breed. Prerequisite for successful Heckel Discus breeding are optimum water conditions. In addition, the dietary regimen for the adults also plays an important role.

Bottom: Discus larvae—a few days old—can be seen here feeding on the skin secretion of their parents. This first food is of fundamental importance for rearing these young fish.

diseases to their progeny. During the first few days the young always remain close to their parents. Parent bonding seems to diminish gradually after two to three weeks. At that point in time the young start swimming around the aquarium and tend to search for larger amounts of food on their own. But even with increasing age they still feed to some degree on the skin secretion of their parents. Parents caring properly for their brood tend to form large amounts of this skin secretion, which often hangs like visible mucus threads from the bodies of the adult fish.

CORRECT REARING METHODS

If proper preparations have been made, getting discus to spawn is not as difficult as is often assumed. Naturally, breeding wild-caught discus is somewhat more complicated, but it is still quite possible. Rearing the young discus successfully presents much larger problems to discus breeders.

Rearing them means caring for them for about eight weeks before the young discus can be sold. During this period the fish are to be reared from the larval stage to a size of about three to four centimeters. If, however, it is intended to raise the discus to adult size, one has to calculate it will take at least one year.

A fundamental prerequisite for good growth are healthy fish. In order to keep the young discus healthy, there must be regular partial water changes, since the more water changes there are, the better it is for the growing fish. Regular partial water changes dilute harmful substances and metabolic waste products which steadily accumulate in an aquarium. Even the most efficient filter alone can not guarantee this. Debris

↑ *Sometimes the adults will eat their eggs or larvae. In order to prevent this, one can place a piece of wire mesh over the eggs. The young will swim through the wire mesh toward the bodies of their parents. Such a protective device is, of course, no cure-all. When parents eat their eggs or larvae, there must have been some sort of disturbance. The adults may in some instances eat their own eggs repeatedly, only to suddenly raise the next lot quite normally. The reason(s) for such behavior has so far not been uncovered.*

← *With increasing age the young will move further and further away from the bodies of their parents. At that point they will search through the entire aquarium for food. An excellent first food is Artemia nauplii. But since these crustaceans have been raised in salt water they should be kept in fresh water for half an hour before they are fed to the young discus. This reduces the salt content in the nauplii and they are more readily digestible by the small discus.*

→ A group of progeny from wild-caught discus from a cross between a Brown and a common Blue Discus. There is a distinct color difference in the brilliant Turquoise Discus in the front of the photograph. The progeny of the wild-caught discus show uniformly lesser but more intensive head markings. This characteristic can be enhanced by way of further selective breeding with the most colorful specimens, which can lead to high quality standards.

and accumulated leftover food must be siphoned off the bottom every day. This by itself already automatically provides a minor partial water change.

Gill and skin damage as well as other external flaws are generally due to inadequate aquarium conditions, most notably poor water quality. For rearing discus they can be acclimated to harder water; in fact, medium hard tap water can be used for that purpose. It is, however, advisable to always filter tap water over activated charcoal before using it in a discus tank.

Discus that have just fed and their gastro-intestinal tract is full of food tend to have a faster respiratory rate and the gill covers are pumping strongly; this is quite normal. Of course, fish like that do not suffer from gill parasites. A specific diet is also very important for successfully rearing discus. Rearing young discus exclusively on one kind of food is not very satisfactory. You should offer your fish a variety of different foods, which also assures the supply of sufficient nutrients and vitamins. Some of the best discus foods include frozen *Artemia* (brine shrimp), turkey heart mixtures, glass worms and mosquito larvae, water fleas (*Daphnia*) and white worms (*Enchytreae*). Generally you should avoid feeding tubifex and bloodworms, although the latter can be offered in order to stimulate newly imported fish to feed in captivity. But later on you should switch over to other types of food.

Ocean Nutrition packages a unique DISCUS FORMULA frozen food which is endorsed by America's leading discus breeder, Jack Wattley.
→

In the juveniles shown below there is already quite a good amount of red coloration. This discus is the progeny of a cross between a Brown and a Green Discus. The strong reddish-brown coloration in the fin margins as well as along the body, is the result of a large red component of the brown ancestors. The extremely high body shape can also be improved through selective breeding. ↓

Once spawning has been completed the parents fan the eggs (1). Although a few individual eggs will always be attacked by fungus, the remaining eggs develop normally (2). After 50 to 60 hours the larvae begin to hatch from the eggs. The first part to appear is the wiggling tail (3).

The yolk sac is then clearly visible, which continues to nourish the young until they are finally swimming freely (4). The young must then swim toward their parents (5). During this phase the parents develop a skin secretion that provides important bacteria and nutrients for the young (6).

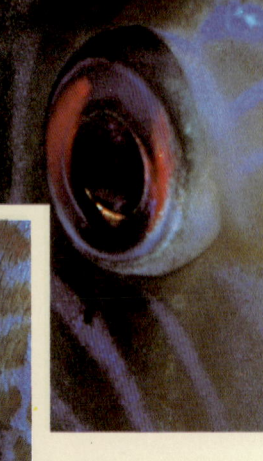

FROM EGG TO JUVENILE FISH

Busily the young graze along the skin of their parents and fill their stomach quickly with the important secretion (7). The formation of this skin secretion turns the adults very dark. Sometimes the secretion is clearly visible (8).

Even at the age of 10 days the young are still feeding continuously from the body of their parents (9). With increasing age the young move more and more away from their parents and will then take all sorts of food, provided it is of the right size (10).

85

Genetics

All discus varieties can be successfully crossbred. This applies to wild-caught specimens as well as to captive-bred stock, so you can cross a Blue Discus with a wild-caught Green Discus without any problems. Moreover, it is also possible to successfully mate a specimen from your Turquoise Discus progeny with a wild-caught Brown Discus. The results of such crosses can sometimes be quite fascinating indeed.

The color of wild-caught discus is the dominant and determinant color when such a fish is crossed with captive–bred stock. In order to appreciate why wild-caught discus have different colors and to find out how to produce breed-specific colors, we have to look at the genetics involved. The laws of genetics have been published as early as 1865 by the Augustine monk Gregor Mendel; however, his theories about genetics were only accepted at the beginning of the 20th century. Genetics deals with the passing on of hereditary traits from one generation to the next and attempts to explain similar features, characteristic traits and colors.

For a beginning aquarist who is involved with intensive, selective breeding of discus for the very first time, the Mendelian laws may not be easy to understand. Let us attempt to find an explanation in a less complicated manner.

The basic female reproductive unit is the egg or ovum, that of the male is referred to as the sperm cell or sperm. As soon as a sperm has entered an egg, a new life develops. Both parents contribute half of the genes. A gene can be compared to a small packet which contains coded information. The genes are connected with each other like a chain of pearls to form a so–called chromosome.

This photograph demonstrates the diversity of colors in discus. All specimens depicted here are the progeny of wild-caught parents. It was selective breeding that intensified the coloration to such a degree. On the left is a typical Brilliant Turquoise Discus, next to that one is an excellently colored Red Turquoise Discus, and again next to that one is a nearly uniform Cobalt Blue Discus. On the righthand side in the back there is a cross variant of an Alenquer Discus.

Pair of wild-caught Green Discus which bonded in a large planted aquarium. With a little bit of luck this discus fancier will be able to breed these fish in a planted aquarium.

Chromosomes always occur in pairs, except in sperm and egg cells. Here the chromosomal pair splits; half of the genes, those from the mother, are in the egg cells; the other half (from the father) are in the sperm cell. Both halves come together again in the fertilized eggs and new life develops. The newly composed pair of chromosomes contains the entire, detailed hereditary material of the new discus fish. But what happens when both halves are different? A young fish can not be simultaneously large and small, blue or green.

Mendel observed that when he crossed garden peas with miniature peas, the progeny was not of medium size, but all were large, that is, the hereditary factor (i.e. the gene) for "large" had persevered. It dominated the hereditary factor "small." This lead to the concept of "dominance" (or being "dominant") in genetics. The weaker hereditary factor "small" dropped into the background which made it in effect "recessive."

Therefore, it often seems as if some hereditary traits skip one generation. That is, they are present but do not appear. Nowadays, we know that that is the way it is. It took someone like Mendel to unravel this secret. When he raised the second generation of peas (those that were large, but were the progeny of large and small peas), he found that not all new peas were large. Every fourth one was small. Even in the third generation there were no medium-size peas. When he crossed the dwarf peas from the third generation among each other he obtained only dwarf peas.

But when crossing the large peas, however, the result was different than expected. Some produced only large peas, others produced dwarf peas as well. Mendel established the following statistical model: 25% dwarf peas, 75% large peas. Among these 75%, only one-third was pure-bred large. The remaining two-thirds were also large, but also carried the recessive trait of "small."

Now we can transpose Mendel's research results to discus. We replace the large peas with green wild-caught discus, and instead of dwarf peas we take blue wild-caught discus. Breeders know from years of experience that the color "green" is dominant over all other colors, and that blue is recessive to green. What happens when the differentiated genes encounter each other? In order to make the answer easier, we call the dominant color "pure green." The combination of dominant green with recessive blue we call "hybrid green." Since blue is recessive, meaning it will be suppressed, there is no "hybrid blue." Therefore, we call it "pure blue". Let us assume, now, that such a pair of discus raises 100 young. The result could be six different combinations.

1. If both parents are "pure green," all of the 100 young would be the same.

2. If both parents are "pure blue," all of the 100 young would be the same.

3. If one parent is "pure green" and the other "pure blue," all 100 young would be "hybrid greens."

A small school of young wild-caught Green Discus, which appear to do well in a planted aquarium. This type of natural discus care is desirable and—with proper quarantine—also quite achievable.

4. If both parents are "hybrid greens," 25 of the young would be "pure green," 25 would be "pure blue," and the remaining 50 young would be "hybrid green."

5. If one parent is "pure green" and the other is "hybrid green," 50 of the young will turn out to be "pure green" and the remaining 50 "hybrid green."

6. If one parent is "pure blue" and the other "hybrid green," 50 of the young will be "pure blue" and the other 50 "hybrid green."

Of course, pure greens and hybrid greens are essentially indistinguishable on the basis of external characteristics. But the pure green discus will produce only pure green progeny, while hybrid green discus progeny will contain green as well as blue young.

These details may be somewhat confusing, but they can provide valuable guidelines when selecting future brood stock.

MUTATIONS

You may ask yourself now how – in view of such a mathematically precise transfer of hereditary material – can there be any development of new colors and other characteristics. And what about evolution? The answer is that somehow, and due to some still unknown process, the genes for color and shape are sometimes

Pigeon Blood Discus; a hereditary mutation from Thailand.

changed in one of the parents, and are passed on in that new form to the next generation. Such a non-repeatable and spontaneous modification is called a mutation. In other words, we are talking about a mutation having taken place when a gene is changed in such a way that a new form or color is passed on. For instance, mutations could have been involved in the totally different discus which have appeared in Asia in recent years.

HYBRIDIZATION

This term must not be confused with mutation. A hybrid is the progeny of two different animal species. An example could be if a discus was crossed with a piranha! This is of course impossible, but there are a lot of valid examples among goldfish varieties and related forms.

In the above examples we have used the term

Magnificent wild-caught Green Discus from Lago Tefe.

Hagen makes a superior flake food for discus. It is called NUTRAFIN and is highly recommended. ←

"hybrid" in order to simplify the explanations of genetically transferring color among discus. The correct scientific term for a discus which has the recessive gene in its genetic make-up is heterozygous. Discus without a recessive gene (for color) are referred to as being homozygous. New discus varieties appear frequently on the market that have new, interesting coloration or markings. These fish have been produced by cross-breeding with other strains or varieties. Since all discus can be cross-bred among one another, it is quite possible to use combinations of different colors and patterns, and from that progeny use selected back-cross combinations to obtain yet further color variations. Of course, rarely do we ever obtain pure-bred discus varieties this way. Most of these new color forms are from ad-hoc crosses. Only through prolonged back-crossing and stabilization of the respective genes is it possible to establish a new discus strain for an extended period of time. Genetic set-backs are always possible in such attempts. Moreover, it pre-supposes the availability of a large number of tanks for raising many hundreds of discus in order to stabilize new color varieties, something that is usually beyond the reach of individual aquarists. Currently, there is a distinct trend toward maintaining the typical wild forms and colors of existing discus varieties. This fact is very encouraging. Let us remember nature and so try to maintain the wild forms in captive breeding activities.

Hybrid of discus progeny with a conspicuous dotted color pattern. This interesting discus is on offer in southeast Asia as the Snakeskin Discus. So far only individual specimens have been available, since this fascinating pattern appears only at random in captive-bred discus. It stands to reason that many more interesting discus variants will reach us from Asia in the future.

CROSS-BRED VARIETIES

A cross variant of the Ghost Discus and a Pigeon Blood Discus by Gan in Singapore. This fish was bred for the first time in Penang/Malaysia.

Cross variant "Red Scribbled" from wild-caught discus from the Rio Purus and Red Turquoise Discus.

Offspring from an Alenquer Discus with brownish-red coloration. With a suitable diet this fish will eventually become intensely red. →

In this Alenquer offspring the red color has not yet fully developed. By increasing the carotene content in the diet the color can be further intensified. Therefore, supplementing the diet with adult brine shrimp is quite advantageous ↓

→ F_3 progeny of Green Discus, which probably came from random stock. Maybe even Brown Discus have already been crossed into this variant. The base body color is already slightly washed out. Body shape and red eye are attractive, but the coloration is not satisfactory.

← Cross of wild-caught Blue Discus with Turquoise progeny. The brownish base coloration has so far been maintained, while the turquoise-colored stripes of the ancestors have been lost. These new progeny now show a pearly or spotted turquoise pattern. These kinds of discus are often sold as Pearl Discus. This term is most commonly used in Asia.

Captive-bred pair derived from the lineage of Brown and Red Turquoise Discus. The combination of strong brown basic tones and the intensive turquoise coloration make these captive-bred fish very appealing discus. These positive traits will be passed on to future generations.

Captive-bred pair of a cross between wild-caught Rio Purus Discus and captive-bred Red Turquoise Discus. These fish will maintain their excellent body shape and interesting coloration over many generations.

Progeny of a mixed cross between Brown and Green Discus. The enormous size of these fish is due to continuous inbreeding, but color was partially sacrificed in the process. These giant discus were difficult to breed. It took a lot of patience and optimum care for these fish to eventually spawn. The female (upper left) normally showed more brown and less turquoise. All males had a greenish base coloration. Some of the males developed the entire color complement, like the specimen on the right.

Variant of a cross between a wild-caught Blue Discus and a captive-bred Red Turquoise Discus. The well-developed finnage and straight vertical bands of this male are striking. Since there is currently a trend among discus fanciers toward the color red, discus with a large red component or a pattern of red lines are very popular.

This breeding pair is the result of an attempted cross between wild-caught Green Discus from Rio Tefe. The pattern of red lines is very delicate and is completely interrupted in the middle of the body, so that numerous dots have developed there. Specimens of this type are commonly bred in southeast Asia. They are popular with aquarists since they are very attractive.

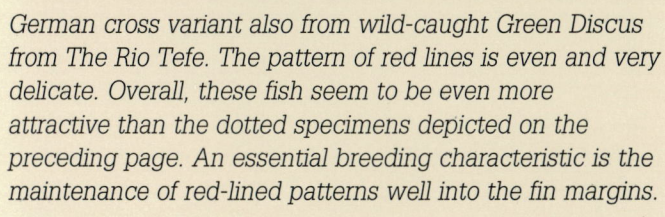

German cross variant also from wild-caught Green Discus from The Rio Tefe. The pattern of red lines is even and very delicate. Overall, these fish seem to be even more attractive than the dotted specimens depicted on the preceding page. An essential breeding characteristic is the maintenance of red-lined patterns well into the fin margins.

A Pearl Turquoise Discus from southeast Asia. Less attractive on this specimen is the small indentation above the snout. This tends to occur in some captive-bred discus stocks; unfortunately, it is genetically inheritable. This fault can only be eliminated through strict selective breeding procedures.

103

Variant cross between wild-caught Blue Discus from the Rio Purus and Red Turquoise Discus. A specimen in a net allows a more detailed color analysis and determination. The intensely dark coloration of the tail fin is conspicuous.

Sibling specimen to the one depicted above. In this specimen the line pattern is already clearly interrupted and dots have started to appear in the middle of the body. It is the breeding objective to achieve further interruption of the stripes so that a pearl-like pattern emerges.

The Thai breeder Kitti Phanaitthi has successfully produced a discus variant. In the Discus Yearbook (1992) there was a report by Kitti about breeding the Golden Rainbow Discus. These fish were later traded as Pigeon Blood Discus. Now Kitti has succeeded for the first time in successfully crossing the Pigeon Blood Discus with Brown Discus, with this variant as the result. However, it is not yet known whether these Red Discus can indeed be bred in large quantities. Yet, with the ambition and determination of the Asians, it can be assumed that this will be possible.

Left top: Variant cross between Blue and common Green Discus. The Coloration appears slightly washed out, but the body shape and eyes are immaculate.

Left bottom: Cross between Heckel Discus and Blue Discus. Coloration and characteristics of the Heckel Discus are still dominant in this specimen.

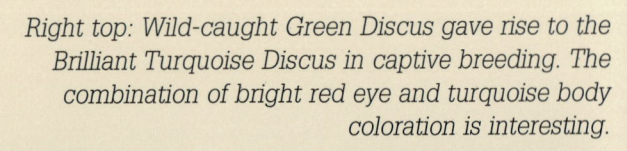

Right top: Wild-caught Green Discus gave rise to the Brilliant Turquoise Discus in captive breeding. The combination of bright red eye and turquoise body coloration is interesting.

Page 106 bottom right: This is a captive-bred specimen from a cross between Green and Brown Discus. The black fin band of the Green Discus against a brownish base coloration is characteristic.

Below: Top-quality specimens from southeast Asia with excellent body form and a beautiful red eye. Specimens of such quality should be used for future breeding. The results should be more than satisfactory.

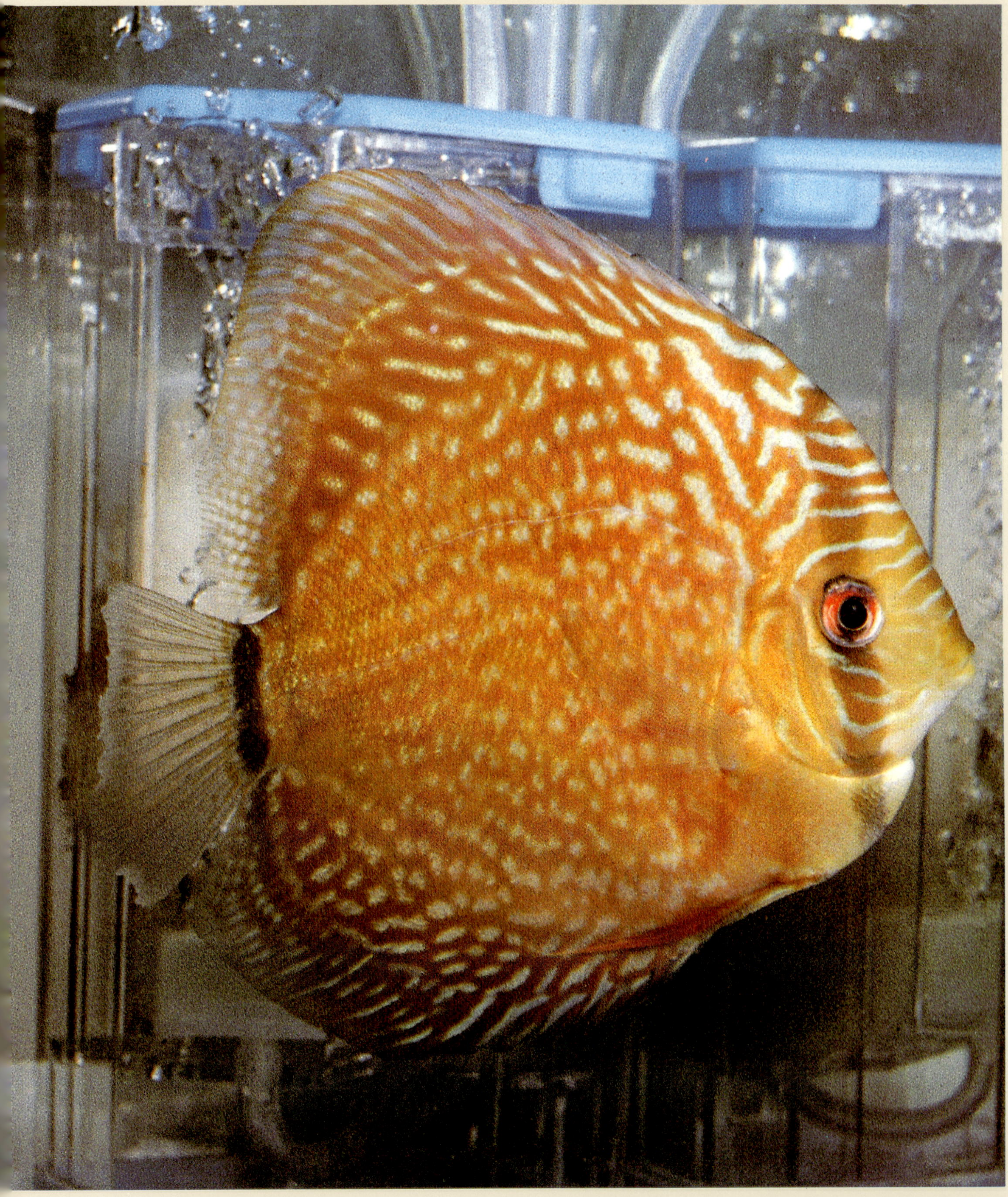

Cross between Brown and Blue Discus. This captive-bred specimen comes from breeding stock which has already been maintained for several generations. The brownish base coloration is sprinkled with turquoise blue dots, which are not yet fully developed. Their color intensity will increase with age. At that point this will be an attractively marked Pearl Discus. ←

↓ *Typical representative of the color variety Red Turquoise. This is a collective term for discus which have red stripes against a turquoise base coloration. Sometimes the red-striped pattern is not very intense, but the fish would still be traded as Red Turquoise. The dorsal fin in this specimen has some beautiful red.*

An interesting Red Turquoise juvenile, which at the age of five months displays surprisingly strong coloration. With appropriate care and a correct diet this fish will grow into a superbly colored specimen. Eye size is ideal relative to body size.

Finally, another wild-caught Green Discus, as commonly traded these days. Would it not be worthwhile again in the future to keep such magnificent specimens in a planted aquarium?

Suggested Reading

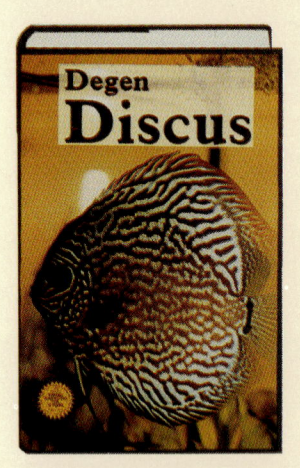

TS-134; 112 pages over 150 full-color photos, 9x12" hard cover.

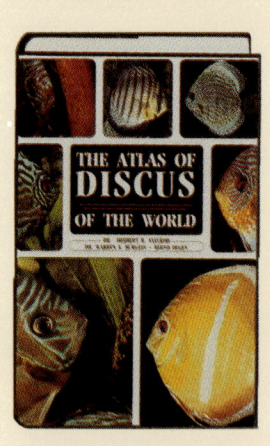

TS-164; 368 pages over 300 full-color photos, 10x13" hard cover.

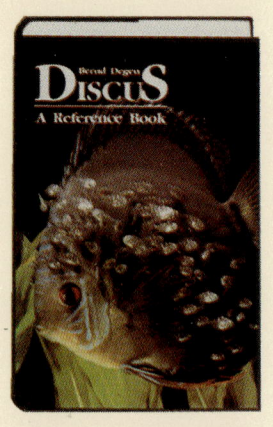

TS-163; 128 pages over 115 full-color photos, 8½x11" hard cover.

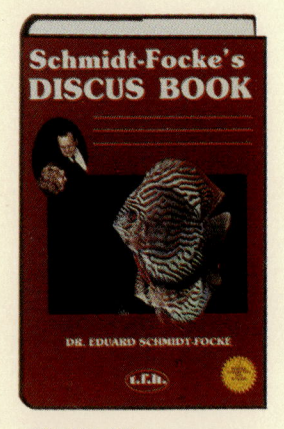

TS-135; 160 pages over 180 full-color photos, 8½x11" hard cover.

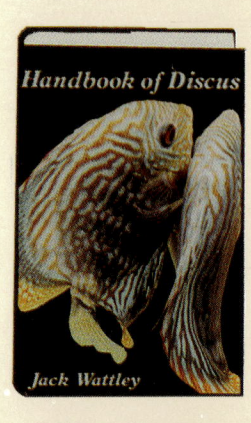

H-1070; 112 pages contains over 100 full-color photos, 8½x11" hard cover.

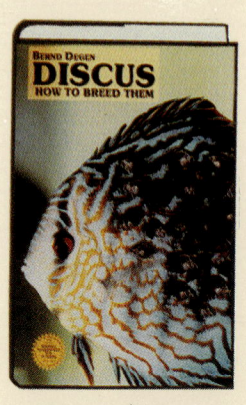

TS-137; 128 pages over 100 full-color photos, 8½x11" hard cover.

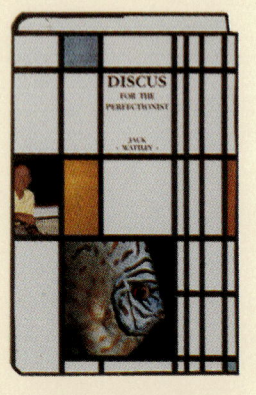

TS-167; 128 pages full-color photos throughout, 8½x11" hard cover.

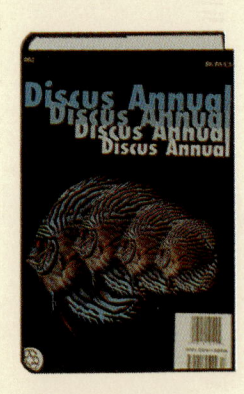

TS-174; 192 pages 8½x11" soft cover.

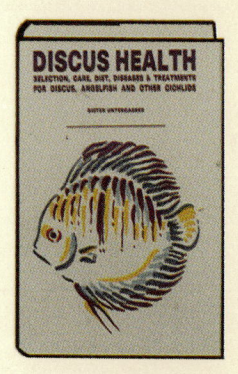

TS-169; 416 pages full-color photos throughout, 7x10" hard cover.

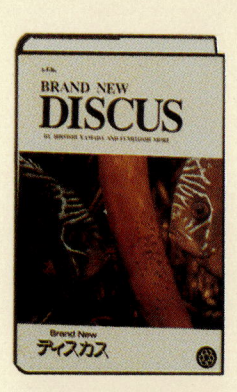

TS-171; 128 pages full-color photos throughout, 7x10" hard cover.

TS-170; 128 pages full-color photos throughout, 7x10" hard cover.

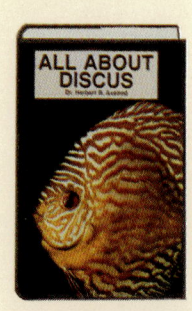

PS-669; 128 pages 95 photos, 5½x8" hard cover.

SK-008; 64 pages 5½x8" soft cover.